MW00355388

We are going to learn all about fractions. First we are going to learn that a fraction has two parts.

NUMERATOR
___
DENOMINATOR

The **NUMERATOR** "numbers" the parts.

The **DENOMINATOR** "names" the parts that make the whole region or set.

As in the fraction:

$\dfrac{7}{8}$

7 numbers the parts

and

8 names the parts of the whole

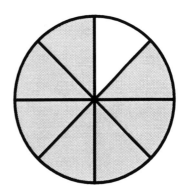

$\dfrac{7}{8}$ parts shaded
parts in the whole region

Write the fraction for the part shaded in each region.

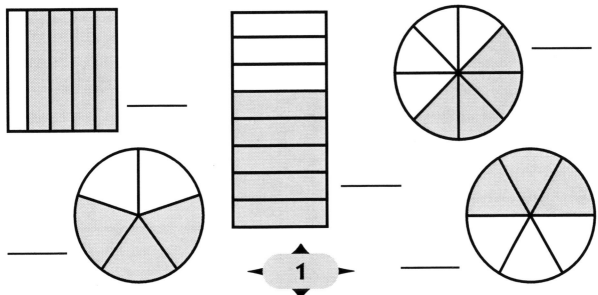

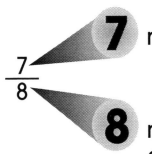

1

FRACTIONS

Write a fraction to show the part shaded in each set.
Follow the example.

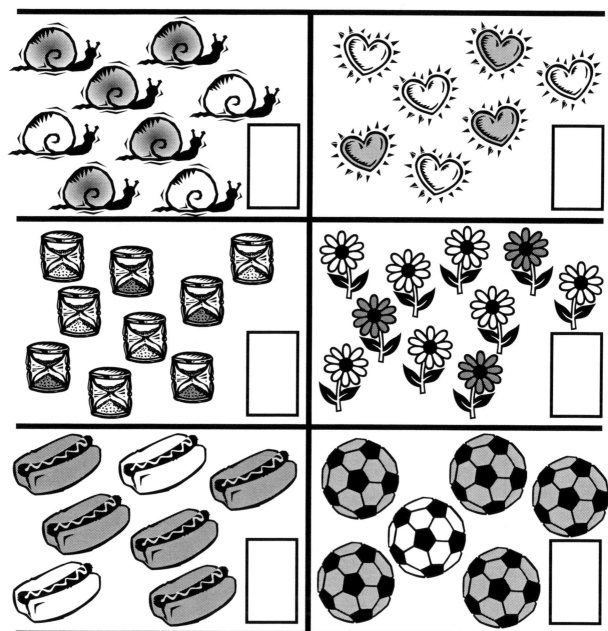

$\dfrac{2}{5}$ parts shaded
parts in the whole

2

SHADED SHAPES.

Color the part of each region **not** shaded.
Then circle the fraction that tells the part you colored.

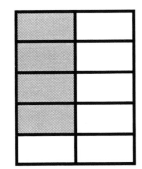

$\frac{5}{10}$ $\frac{6}{8}$ $\frac{6}{10}$

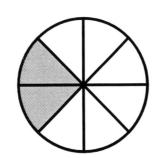

$\frac{2}{8}$ $\frac{6}{8}$ $\frac{8}{10}$

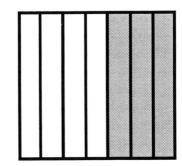

$\frac{3}{7}$ $\frac{4}{7}$ $\frac{4}{8}$

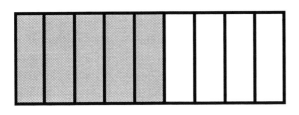

$\frac{4}{9}$ $\frac{4}{8}$ $\frac{5}{9}$

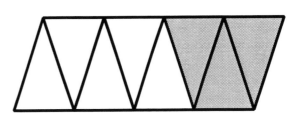

$\frac{1}{2}$ $\frac{4}{8}$ $\frac{5}{8}$

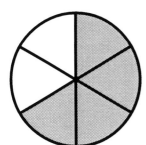

$\frac{1}{2}$ $\frac{2}{6}$ $\frac{4}{6}$

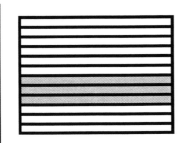

$\frac{7}{10}$ $\frac{7}{12}$ $\frac{9}{12}$

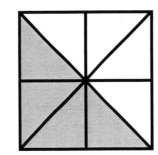

$\frac{4}{8}$ $\frac{4}{6}$ $\frac{6}{8}$

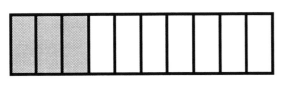

$\frac{3}{9}$ $\frac{3}{10}$ $\frac{7}{10}$

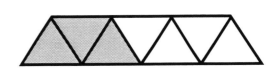

$\frac{2}{7}$ $\frac{4}{7}$ $\frac{5}{7}$

◀ 3 ▶

Write the fraction to tell the part of each region that is shaded. Then use the code to find the answer to the riddle.

T	O	S	L	E	I	M
$\frac{1}{3}$	$\frac{2}{5}$	$\frac{3}{5}$	$\frac{3}{4}$	$\frac{1}{2}$	$\frac{2}{3}$	$\frac{1}{4}$

If an athlete gets athlete's foot, what does an astronaut get?

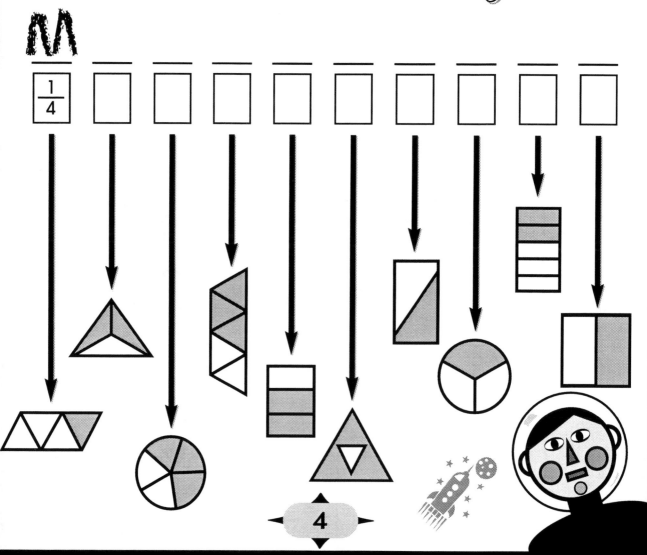

M
$\frac{1}{4}$

4

These objects are divided into equal parts.
They are the **DENOMINATORS**.
Write the denominator of each object.

 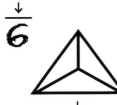 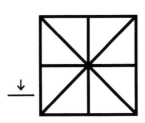  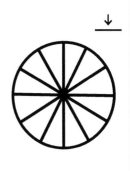

The **NUMERATOR** is the <u>number</u> of parts we are describing.
In the shapes below, it is the colored part. Write the numerator
for each object.

 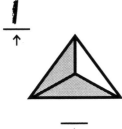 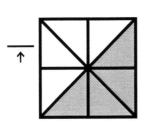  

Write the **NUMERATOR** and **DENOMINATOR** for each object below.

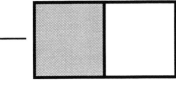 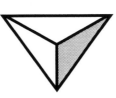 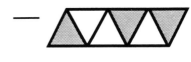

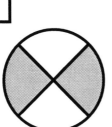

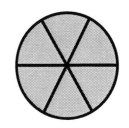

 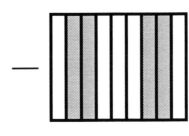

— or

FRACTIONS

Write the fraction to tell the part that is shaded.
Compare the fractions, then put **<**, **=**, or **>** in the circle.

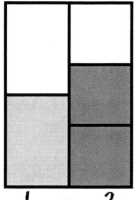

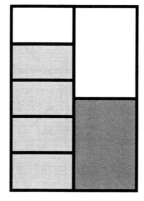

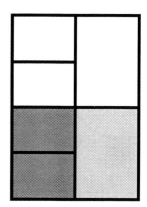

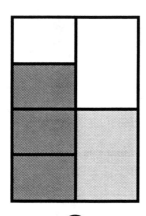

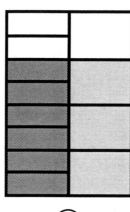

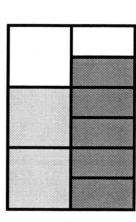

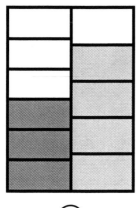

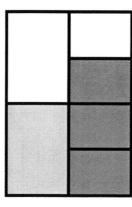

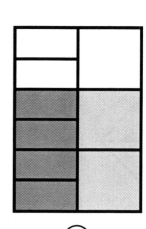

Two-fourths **EQUALS** one-half.
They are **EQUIVALENT** fractions.
The same amount is shaded in both regions.

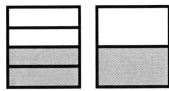

$$\frac{2}{4} = \frac{1}{2}$$

Write the **EQUIVALENT FRACTION** in each pair of regions.

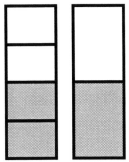

$$\frac{2}{4} = \underline{}$$

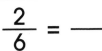

$$\frac{2}{6} = \underline{}$$

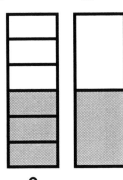

$$\frac{3}{6} = \underline{}$$

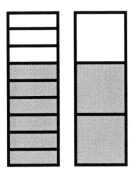

$$\frac{6}{9} = \underline{}$$

$$\frac{6}{8} = \underline{}$$

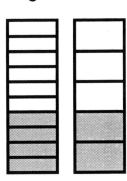

$$\frac{4}{10} = \underline{}$$

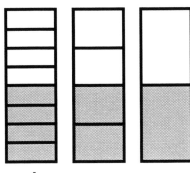

$$\frac{4}{8} = \underline{} = \underline{}$$

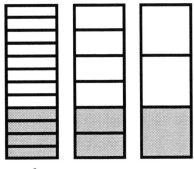

$$\frac{4}{12} = \underline{} = \underline{}$$

7

Shade fractions equivalent to $\frac{1}{2}$.

Other names for $\frac{1}{2}$:

$\frac{2}{4}$, $\frac{3}{6}$, $\frac{4}{8}$, etc.

$\frac{10}{32}$ $\frac{9}{11}$ $\frac{7}{21}$ $\frac{8}{11}$ $\frac{7}{10}$

$\frac{14}{30}$ $\frac{7}{9}$ $\frac{12}{36}$ $\frac{10}{50}$

$\frac{5}{7}$ $\frac{2}{14}$ $\frac{8}{24}$ $\frac{7}{15}$

$\frac{12}{32}$ $\frac{1}{6}$ $\frac{1}{4}$ $\frac{4}{11}$ $\frac{3}{8}$ $\frac{4}{8}$ $\frac{1}{2}$ $\frac{3}{6}$ $\frac{1}{8}$ $\frac{24}{30}$ $\frac{20}{50}$

$\frac{5}{15}$ $\frac{3}{5}$ $\frac{4}{6}$ $\frac{6}{10}$ $\frac{6}{12}$ $\frac{5}{10}$ $\frac{5}{9}$ $\frac{12}{20}$ $\frac{24}{50}$ $\frac{20}{36}$

$\frac{18}{30}$ $\frac{12}{24}$ $\frac{10}{20}$ $\frac{2}{4}$ $\frac{7}{14}$ $\frac{11}{22}$ $\frac{8}{16}$ $\frac{8}{15}$ $\frac{12}{30}$ $\frac{10}{12}$

$\frac{2}{4}$ $\frac{21}{42}$ $\frac{14}{28}$ $\frac{3}{6}$ $\frac{6}{8}$ $\frac{22}{44}$

$\frac{3}{9}$ $\frac{2}{9}$ $\frac{3}{6}$ $\frac{10}{24}$ $\frac{2}{3}$ $\frac{26}{52}$ $\frac{10}{16}$ $\frac{4}{24}$

$\frac{5}{20}$ $\frac{5}{10}$ $\frac{4}{8}$ $\frac{9}{18}$ $\frac{12}{24}$ $\frac{3}{4}$ $\frac{17}{20}$ $\frac{1}{2}$ $\frac{4}{9}$ $\frac{49}{98}$

$\frac{8}{36}$ $\frac{7}{14}$ $\frac{6}{12}$ $\frac{13}{26}$ $\frac{4}{8}$ $\frac{4}{10}$ $\frac{12}{18}$ $\frac{5}{8}$ $\frac{7}{10}$ $\frac{6}{7}$

$\frac{3}{8}$ $\frac{2}{7}$ $\frac{25}{50}$ $\frac{16}{24}$ $\frac{24}{48}$ $\frac{9}{12}$ $\frac{4}{7}$ $\frac{8}{12}$ $\frac{12}{14}$ $\frac{2}{8}$

$\frac{30}{60}$ $\frac{7}{16}$ $\frac{2}{10}$

$\frac{15}{30}$ $\frac{8}{16}$ $\frac{6}{12}$ $\frac{6}{14}$ $\frac{6}{18}$

$\frac{8}{9}$ $\frac{17}{34}$ $\frac{15}{20}$ $\frac{9}{18}$ $\frac{15}{25}$ $\frac{20}{30}$

$\frac{10}{14}$ $\frac{3}{12}$ $\frac{5}{10}$ $\frac{35}{70}$ $\frac{2}{6}$ $\frac{10}{25}$ $\frac{4}{9}$ $\frac{1}{3}$

$\frac{18}{20}$ $\frac{27}{54}$ $\frac{8}{10}$ $\frac{6}{9}$ $\frac{5}{11}$

$\frac{5}{12}$ $\frac{16}{32}$ $\frac{2}{4}$ $\frac{18}{36}$ $\frac{7}{9}$

$\frac{20}{40}$ $\frac{9}{14}$ $\frac{3}{10}$ $\frac{3}{7}$ $\frac{18}{24}$ $\frac{12}{25}$

$\frac{19}{38}$ $\frac{8}{14}$ $\frac{1}{4}$ $\frac{14}{24}$ $\frac{5}{25}$

$\frac{7}{11}$ $\frac{9}{10}$ $\frac{7}{8}$

$\frac{20}{24}$ $\frac{16}{30}$ $\frac{1}{4}$ $\frac{6}{15}$ $\frac{2}{5}$ $\frac{10}{30}$ $\frac{5}{12}$ $\frac{5}{6}$ $\frac{8}{20}$

$\frac{5}{6}$

8

EQUIVALENT FRACTIONS.

Match to show equivalent fractions.

FRACTIONS

EQUIVALENT FRACTIONS.

Color in one part of the graph to show an equivalent fraction for each of the shaded regions below.

 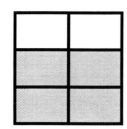 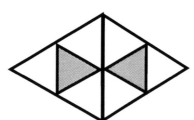 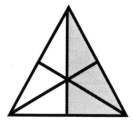

___ ___ ___ ___

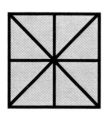 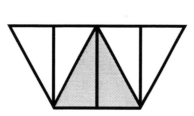 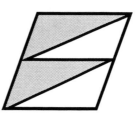 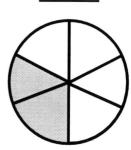

___ ___ ___ ___

Look at the graph.

Which fraction was used most?

___

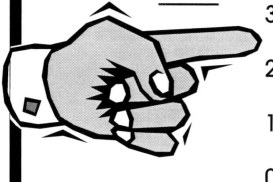

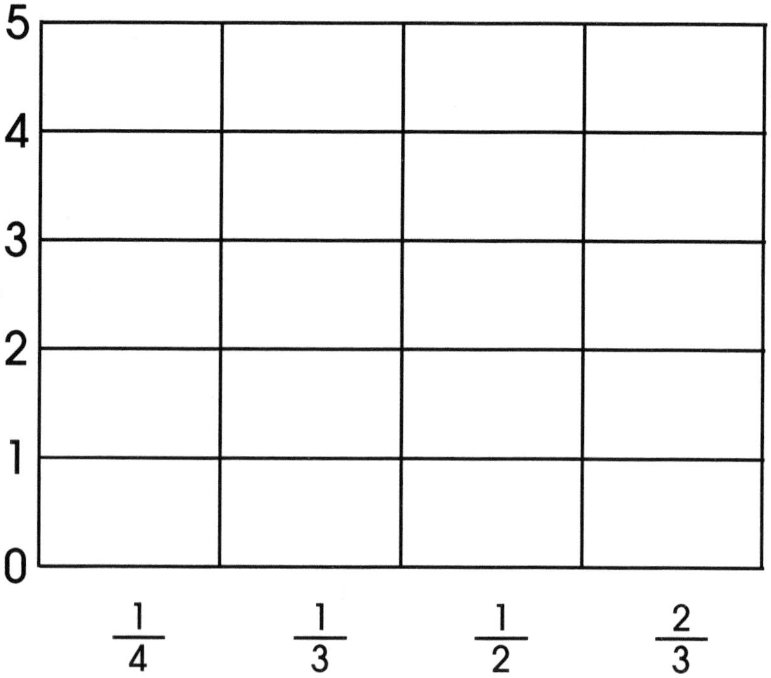

Graph with vertical axis labeled 0, 1, 2, 3, 4, 5 and horizontal axis labeled $\frac{1}{4}$, $\frac{1}{3}$, $\frac{1}{2}$, $\frac{2}{3}$.

Shade fractions equivalent to $\frac{1}{4}$.

Other names for $\frac{1}{4}$:

$$\frac{1}{4}, \frac{2}{8}, \frac{3}{12}, \frac{4}{16}, \text{etc.}$$

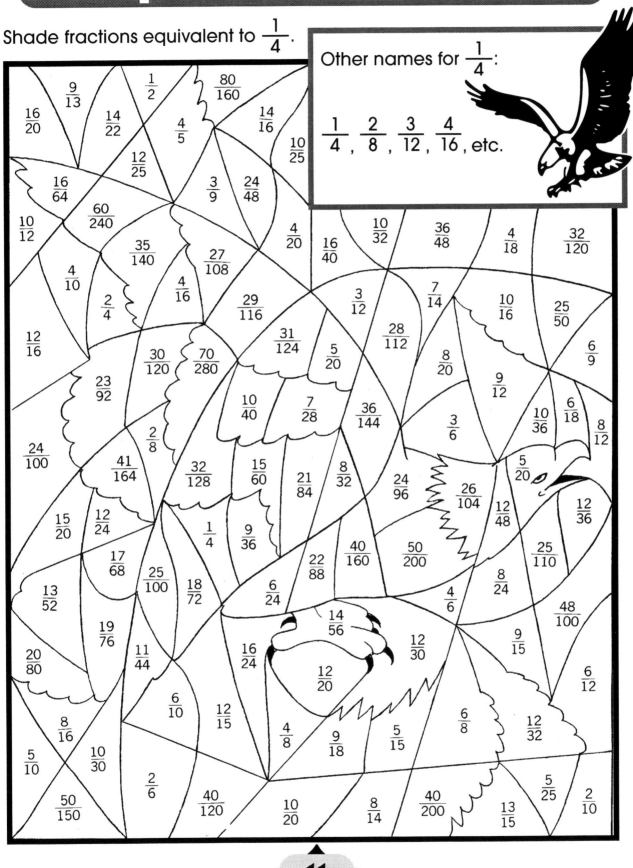

Line AB is divided into **HALVES**.

Line CD is divided into **THIRDS**.

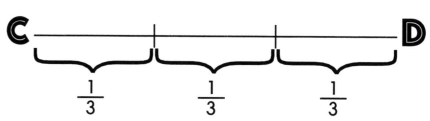

Line EF is divided into **FOURTHS**.

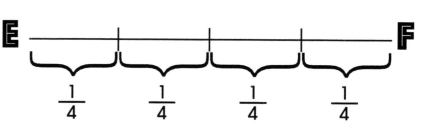

Look at these number lines. Each line is divided into segments. Write the correct fraction in each box.

The segment between 0 and 1 is divided into **halves**.

$\frac{2}{2}$

The segment between 0 and 1 is divided into **thirds**.

$\frac{3}{3}$

The segment between 0 and 1 is divided into **fourths**.

$\frac{4}{4}$

12

MEASURING FRACTIONS.

This is a ruler. It is divided into segments called inches. An **INCH** is a unit used for measuring length.

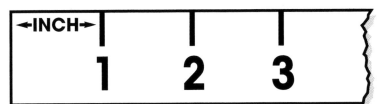

To measure more accurately, this ruler is divided into **HALF INCHES**.

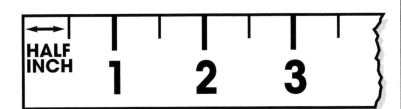

About how long is each line segment?

_____ inches

_____ inches

_____ inch

_____ inches

_____ inches

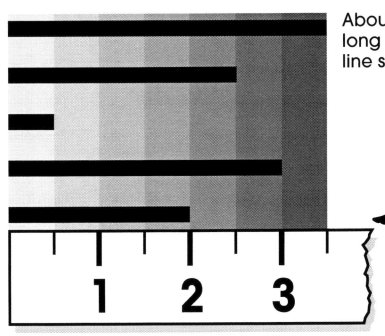

This ruler is divided into fourths.

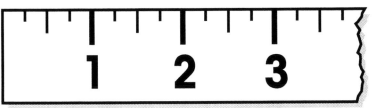

How many fourths in 1 inch? _____

in $\frac{1}{2}$ inch? _____

in 3 inches? _____

in $1\frac{1}{2}$ inches? _____

in $2\frac{1}{2}$ inches? _____

FRACTIONS

Shade fractions equivalent to $\frac{1}{5}$.

Other names for $\frac{1}{5}$:

$$\frac{1}{5}, \frac{2}{10}, \frac{3}{15}, \frac{4}{20}, \text{etc.}$$

$\frac{3}{6}$ $\frac{3}{4}$ $\frac{25}{45}$ $\frac{30}{60}$ $\frac{3}{9}$ $\frac{20}{45}$ $\frac{8}{10}$ $\frac{2}{5}$

$\frac{10}{15}$ $\frac{3}{15}$ $\frac{6}{9}$

$\frac{4}{5}$ $\frac{5}{6}$

$\frac{16}{20}$ $\frac{9}{15}$ $\frac{20}{35}$ $\frac{22}{110}$ $\frac{15}{75}$ $\frac{5}{35}$ $\frac{20}{40}$ $\frac{15}{25}$

$\frac{8}{12}$ $\frac{20}{25}$ $\frac{5}{25}$ $\frac{1}{5}$ $\frac{6}{10}$ $\frac{2}{3}$

$\frac{1}{3}$ $\frac{6}{8}$ $\frac{15}{20}$

$\frac{40}{55}$ $\frac{18}{30}$ $\frac{10}{50}$ $\frac{6}{30}$ $\frac{5}{20}$ $\frac{6}{12}$

$\frac{6}{10}$ $\frac{3}{15}$ $\frac{17}{85}$ $\frac{19}{95}$ $\frac{12}{60}$ $\frac{12}{40}$ $\frac{25}{45}$

$\frac{8}{12}$ $\frac{2}{6}$ $\frac{2}{10}$ $\frac{3}{9}$ $\frac{7}{35}$ $\frac{4}{20}$ $\frac{2}{10}$ $\frac{5}{15}$ $\frac{4}{6}$

$\frac{3}{9}$ $\frac{10}{15}$ $\frac{21}{105}$ $\frac{18}{90}$ $\frac{8}{30}$ $\frac{20}{50}$

$\frac{3}{12}$ $\frac{10}{25}$

$\frac{15}{30}$ $\frac{3}{6}$ $\frac{3}{5}$ $\frac{20}{60}$ $\frac{30}{150}$ $\frac{11}{55}$ $\frac{4}{8}$

$\frac{80}{160}$

$\frac{5}{10}$ $\frac{10}{40}$ $\frac{13}{65}$ $\frac{10}{35}$ $\frac{14}{70}$ $\frac{1}{2}$

$\frac{30}{50}$ $\frac{15}{45}$ $\frac{2}{40}$ $\frac{2}{3}$

$\frac{2}{40}$ $\frac{8}{20}$ $\frac{8}{40}$ $\frac{10}{45}$ $\frac{2}{5}$ $\frac{16}{80}$ $\frac{5}{8}$ $\frac{40}{50}$

$\frac{4}{12}$ $\frac{1}{4}$

$\frac{12}{15}$ $\frac{23}{115}$ $\frac{6}{20}$ $\frac{16}{30}$ $\frac{15}{55}$ $\frac{9}{45}$ $\frac{1}{6}$

$\frac{5}{35}$

$\frac{3}{8}$ $\frac{20}{60}$ $\frac{2}{4}$ $\frac{15}{40}$ $\frac{9}{12}$ $\frac{5}{25}$

$\frac{20}{100}$

$\frac{10}{20}$ $\frac{10}{40}$ $\frac{50}{55}$ $\frac{5}{10}$

EQUIVALENT FRACTIONS.

Use this chart and a ruler to find equivalent fractions.

1											
$\frac{1}{2}$						$\frac{1}{2}$					
$\frac{1}{3}$				$\frac{1}{3}$				$\frac{1}{3}$			
$\frac{1}{4}$			$\frac{1}{4}$			$\frac{1}{4}$			$\frac{1}{4}$		
$\frac{1}{5}$		$\frac{1}{5}$		$\frac{1}{5}$		$\frac{1}{5}$			$\frac{1}{5}$		
$\frac{1}{6}$		$\frac{1}{6}$		$\frac{1}{6}$		$\frac{1}{6}$		$\frac{1}{6}$		$\frac{1}{6}$	
$\frac{1}{8}$	$\frac{1}{8}$	$\frac{1}{8}$	$\frac{1}{8}$	$\frac{1}{8}$	$\frac{1}{8}$	$\frac{1}{8}$	$\frac{1}{8}$				
$\frac{1}{10}$	$\frac{1}{10}$	$\frac{1}{10}$	$\frac{1}{10}$	$\frac{1}{10}$	$\frac{1}{10}$	$\frac{1}{10}$	$\frac{1}{10}$	$\frac{1}{10}$	$\frac{1}{10}$		
$\frac{1}{12}$	$\frac{1}{12}$	$\frac{1}{12}$	$\frac{1}{12}$	$\frac{1}{12}$	$\frac{1}{12}$	$\frac{1}{12}$	$\frac{1}{12}$	$\frac{1}{12}$	$\frac{1}{12}$	$\frac{1}{12}$	$\frac{1}{12}$

$$\frac{2}{3} = \frac{}{6} \qquad \frac{1}{5} = \frac{}{10}$$

$$\frac{2}{5} = \frac{}{10} \qquad \frac{1}{2} = \frac{}{4}$$

$$\frac{3}{6} = \frac{}{12} \qquad \frac{1}{3} = \frac{}{6} \qquad \frac{3}{4} = \frac{}{12} \qquad \frac{4}{8} = \frac{}{12}$$

The fractions in these problems are equivalent. Write each fraction.

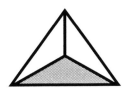 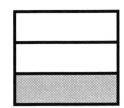  

___   ___   ___   ___

 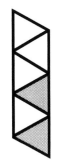 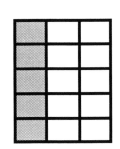 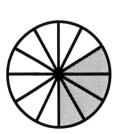

___   ___   ___   ___

15

FRACTIONS

EQUIVALENT FRACTIONS.

EXAMPLE:

$$\frac{2}{7} = \frac{2}{7} \times \frac{3}{3} = \frac{6}{21}$$

You can multiply the numerator and denominator by the same number, except zero, to get an equivalent fraction.

Find equivalent fractions by multiplying each fraction by the colored fraction shown.

$$\frac{1}{3} = \frac{1}{3} \times \frac{2}{2} = \underline{\qquad} \qquad \frac{4}{5} = \underline{\qquad} \qquad \frac{3}{10} = \underline{\qquad} \qquad \frac{1}{6} = \underline{\qquad}$$

$$\frac{7}{10} = \frac{7}{10} \times \frac{3}{3} = \underline{\qquad} \qquad \frac{2}{3} = \underline{\qquad} \qquad \frac{1}{4} = \underline{\qquad} \qquad \frac{1}{5} = \underline{\qquad}$$

$$\frac{3}{8} = \frac{3}{8} \times \frac{4}{4} = \underline{\qquad} \qquad \frac{9}{10} = \underline{\qquad} \qquad \frac{1}{3} = \underline{\qquad} \qquad \frac{7}{8} = \underline{\qquad}$$

Find the **lowest term** fraction. Divide the numerator and the denominator by the largest possible factor.

EXAMPLE:

$$\frac{12}{16} = \frac{12}{16} \div \frac{4}{4} = \frac{3}{4}$$

$$\frac{6}{10} = \frac{6}{10} \div \frac{2}{2} = \frac{3}{5}$$

$$\frac{12}{20} = \underline{\qquad} \qquad \frac{6}{24} = \underline{\qquad}$$

$$\frac{14}{21} = \underline{\qquad} \qquad \frac{14}{16} = \underline{\qquad} \qquad \frac{9}{30} = \underline{\qquad}$$

$$\frac{9}{12} = \underline{\qquad} \qquad \frac{12}{15} = \underline{\qquad} \qquad \frac{7}{21} = \underline{\qquad} \qquad \frac{10}{12} = \underline{\qquad}$$

$$\frac{4}{20} = \underline{\qquad} \qquad \frac{11}{33} = \underline{\qquad} \qquad \frac{5}{30} = \underline{\qquad} \qquad \frac{6}{40} = \underline{\qquad}$$

$$\frac{8}{48} = \underline{\qquad} \qquad \frac{10}{25} = \underline{\qquad} \qquad \frac{3}{27} = \underline{\qquad} \qquad \frac{12}{36} = \underline{\qquad}$$

16

Shade fractions equivalent to $\frac{2}{3}$.

Other names for $\frac{2}{3}$:

$$\frac{2}{3}, \frac{4}{6}, \frac{6}{9}, \frac{8}{12}, \text{etc.}$$

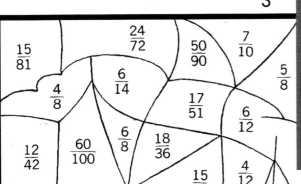

$\frac{15}{81}$ $\frac{24}{72}$ $\frac{50}{90}$ $\frac{7}{10}$ $\frac{5}{8}$
$\frac{4}{8}$ $\frac{6}{14}$ $\frac{17}{51}$ $\frac{6}{12}$
$\frac{12}{42}$ $\frac{60}{100}$ $\frac{6}{8}$ $\frac{18}{36}$ $\frac{15}{75}$ $\frac{4}{12}$ $\frac{25}{45}$ $\frac{3}{12}$ $\frac{8}{14}$
$\frac{9}{24}$ $\frac{12}{45}$ $\frac{30}{60}$ $\frac{15}{20}$ $\frac{46}{69}$ $\frac{34}{51}$ $\frac{27}{54}$ $\frac{50}{100}$ $\frac{12}{45}$
$\frac{7}{12}$ $\frac{54}{81}$
$\frac{15}{48}$ $\frac{6}{10}$ $\frac{30}{48}$ $\frac{10}{18}$ $\frac{3}{9}$ $\frac{20}{100}$
$\frac{12}{36}$ $\frac{75}{100}$ $\frac{16}{24}$ $\frac{40}{60}$ $\frac{11}{33}$
$\frac{15}{27}$ $\frac{12}{15}$ $\frac{52}{78}$ $\frac{15}{48}$
$\frac{3}{8}$ $\frac{8}{16}$ $\frac{30}{45}$ $\frac{12}{36}$ $\frac{4}{16}$ $\frac{12}{30}$
$\frac{2}{4}$ $\frac{4}{6}$ $\frac{9}{12}$
$\frac{10}{15}$ $\frac{15}{45}$ $\frac{24}{36}$ $\frac{48}{72}$ $\frac{44}{66}$ $\frac{15}{60}$ $\frac{3}{6}$ $\frac{14}{32}$
$\frac{14}{42}$ $\frac{7}{21}$ $\frac{28}{42}$ $\frac{50}{75}$ $\frac{42}{63}$ $\frac{32}{48}$ $\frac{6}{15}$ $\frac{5}{6}$
$\frac{4}{5}$ $\frac{20}{60}$ $\frac{200}{300}$ $\frac{38}{57}$ $\frac{100}{150}$ $\frac{9}{27}$ $\frac{20}{30}$
$\frac{36}{54}$ $\frac{8}{12}$ $\frac{4}{10}$
$\frac{10}{14}$ $\frac{14}{21}$ $\frac{8}{24}$
$\frac{3}{4}$ $\frac{6}{12}$
$\frac{10}{42}$ $\frac{60}{75}$ $\frac{80}{120}$ $\frac{9}{10}$ $\frac{2}{8}$ $\frac{30}{75}$
$\frac{10}{15}$ $\frac{16}{18}$ $\frac{3}{5}$
$\frac{22}{33}$ $\frac{26}{39}$
$\frac{18}{20}$ $\frac{18}{27}$ $\frac{6}{9}$ $\frac{9}{15}$
$\frac{7}{8}$
$\frac{9}{36}$ $\frac{10}{25}$
$\frac{7}{14}$ $\frac{5}{10}$ $\frac{10}{45}$ $\frac{3}{10}$ $\frac{15}{36}$ $\frac{8}{10}$

FRACTIONS

Write each fraction in its simplest form.

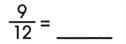

EXAMPLE:

$$\frac{9}{12} = \frac{9}{12} \div \frac{3}{3} =$$

$\dfrac{9}{12} =$ _____ $\dfrac{5}{10} =$ _____ $\dfrac{4}{16} =$ _____

$\dfrac{4}{20} =$ _____ $\dfrac{4}{6} =$ _____ $\dfrac{4}{8} =$ _____

$\dfrac{6}{9} =$ _____ $\dfrac{5}{15} =$ _____ $\dfrac{8}{12} =$ _____

$\dfrac{16}{18} =$ _____ $\dfrac{7}{14} =$ _____ $\dfrac{15}{21} =$ _____ $\dfrac{3}{18} =$ _____

$\dfrac{14}{16} =$ _____ $\dfrac{9}{21} =$ _____ $\dfrac{8}{28} =$ _____ $\dfrac{20}{28} =$ _____

$\dfrac{12}{36} =$ _____ $\dfrac{30}{36} =$ _____ $\dfrac{10}{55} =$ _____ $\dfrac{50}{75} =$ _____

$\dfrac{20}{72} =$ _____ $\dfrac{18}{48} =$ _____ $\dfrac{15}{18} =$ _____ $\dfrac{32}{36} =$ _____

$\dfrac{12}{54} =$ _____ $\dfrac{30}{55} =$ _____ $\dfrac{60}{90} =$ _____ $\dfrac{36}{90} =$ _____

$\dfrac{36}{48} =$ _____ $\dfrac{20}{32} =$ _____ $\dfrac{14}{42} =$ _____ $\dfrac{28}{32} =$ _____

$\dfrac{12}{45} =$ _____ $\dfrac{20}{60} =$ _____ $\dfrac{35}{50} =$ _____ $\dfrac{40}{48} =$ _____

$\dfrac{21}{36} =$ _____ $\dfrac{33}{54} =$ _____ $\dfrac{35}{42} =$ _____ $\dfrac{30}{48} =$ _____

18

EXAMPLE:

$$\frac{1 \times 2}{2 \times 2} = \frac{2}{4} \qquad \frac{1 \times 3}{2 \times 3} = \frac{3}{6}$$

Complete these exercises.

$\dfrac{1}{2} = \dfrac{}{4}$ $\dfrac{1}{2} = \dfrac{}{6}$

$\dfrac{1}{3} = \dfrac{}{9}$ $\dfrac{1}{4} = \dfrac{}{8}$

$\dfrac{2}{5} = \dfrac{}{10}$ $\dfrac{2}{3} = \dfrac{}{6}$ $\dfrac{3}{4} = \dfrac{}{12}$ $\dfrac{5}{6} = \dfrac{}{18}$

$\dfrac{3}{8} = \dfrac{}{16}$ $\dfrac{1}{6} = \dfrac{5}{}$ $\dfrac{1}{8} = \dfrac{4}{}$ $\dfrac{2}{9} = \dfrac{6}{}$

$\dfrac{3}{5} = \dfrac{12}{}$ $\dfrac{5}{12} = \dfrac{10}{}$ $\dfrac{7}{8} = \dfrac{}{24}$ $\dfrac{4}{9} = \dfrac{16}{}$

$\dfrac{4}{5} = \dfrac{}{20}$ $\dfrac{11}{12} = \dfrac{}{36}$ $\dfrac{9}{10} = \dfrac{36}{}$ $\dfrac{7}{8} = \dfrac{21}{}$

$\dfrac{1}{10}, \ \dfrac{}{20}, \ \dfrac{}{30}, \ \dfrac{}{40}, \ \dfrac{}{50}, \ \dfrac{}{60}$ $\dfrac{1}{9}, \ \dfrac{2}{}, \ \dfrac{3}{}, \ \dfrac{4}{}, \ \dfrac{5}{}, \ \dfrac{6}{}$

$\dfrac{2}{7}, \ \dfrac{}{14}, \ \dfrac{}{21}, \ \dfrac{}{28}, \ \dfrac{}{35}, \ \dfrac{}{42}$ $\dfrac{7}{12}, \ \dfrac{14}{}, \ \dfrac{21}{}, \ \dfrac{28}{}, \ \dfrac{35}{}, \ \dfrac{42}{}$

$\dfrac{1}{5}, \ \dfrac{}{10}, \ \dfrac{}{15}, \ \dfrac{}{20}, \ \dfrac{}{25}, \ \dfrac{}{30}$ $\dfrac{3}{10}, \ \dfrac{6}{}, \ \dfrac{9}{}, \ \dfrac{12}{}, \ \dfrac{15}{}, \ \dfrac{18}{}$

$\dfrac{5}{9}, \ \dfrac{}{18}, \ \dfrac{}{27}, \ \dfrac{}{36}, \ \dfrac{}{45}, \ \dfrac{}{54}$ $\dfrac{4}{7}, \ \dfrac{8}{}, \ \dfrac{}{21}, \ \dfrac{}{28}, \ \dfrac{20}{}, \ \dfrac{}{42}$

$\dfrac{1}{12}, \ \dfrac{}{24}, \ \dfrac{}{36}, \ \dfrac{}{48}, \ \dfrac{}{60}, \ \dfrac{}{72}$ $\dfrac{5}{8}, \ \dfrac{}{16}, \ \dfrac{15}{}, \ \dfrac{}{32}, \ \dfrac{25}{}, \ \dfrac{}{48}$

      FRACTIONS

Shade fractions in their simplest form.

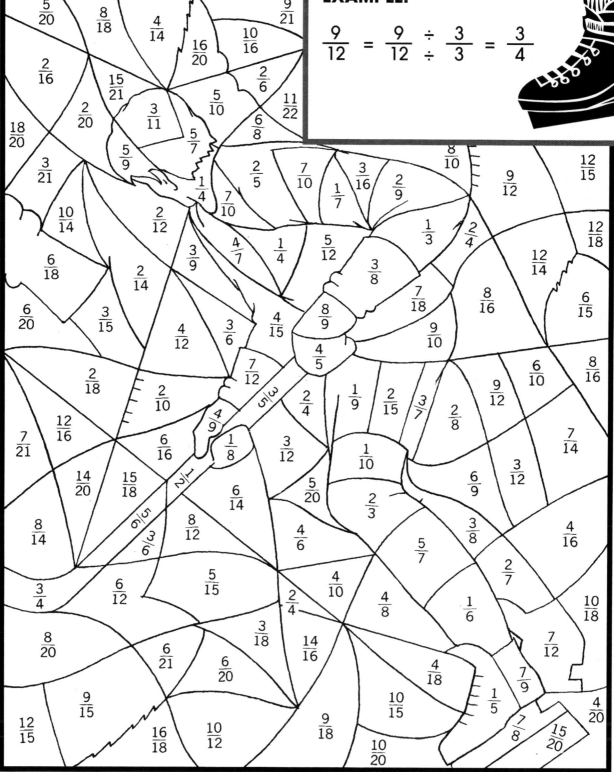

EXAMPLE:

$$\frac{9}{12} = \frac{9}{12} \div \frac{3}{3} = \frac{3}{4}$$

MIXED NUMERALS.

Write each mixed numeral as a fraction.

$$2\frac{2}{5} = \frac{2}{1} + \frac{2}{5} =$$
$$\frac{10}{5} + \frac{2}{5} = \frac{12}{5}$$

$2\frac{2}{5} = \underline{\hspace{1cm}}$ $3\frac{1}{8} = \underline{\hspace{1cm}}$

$1\frac{1}{6} = \underline{\hspace{1cm}}$ $4\frac{1}{4} = \underline{\hspace{1cm}}$ $3\frac{1}{5} = \underline{\hspace{1cm}}$

$2\frac{2}{3} = \underline{\hspace{1cm}}$ $5\frac{8}{9} = \underline{\hspace{1cm}}$ $4\frac{5}{8} = \underline{\hspace{1cm}}$

$5\frac{1}{3} = \underline{\hspace{1cm}}$ $4\frac{1}{2} = \underline{\hspace{1cm}}$ $3\frac{1}{7} = \underline{\hspace{1cm}}$

$3\frac{3}{4} = \underline{\hspace{1cm}}$ $2\frac{5}{6} = \underline{\hspace{1cm}}$ $2\frac{4}{5} = \underline{\hspace{1cm}}$

$2\frac{1}{10} = \underline{\hspace{1cm}}$ $1\frac{4}{9} = \underline{\hspace{1cm}}$ $1\frac{9}{16} = \underline{\hspace{1cm}}$

$1\frac{7}{12} = \underline{\hspace{1cm}}$ $2\frac{4}{7} = \underline{\hspace{1cm}}$ $1\frac{3}{8} = \underline{\hspace{1cm}}$

$6\frac{1}{9} = \underline{\hspace{1cm}}$ $4\frac{3}{5} = \underline{\hspace{1cm}}$ $3\frac{9}{10} = \underline{\hspace{1cm}}$

$2\frac{5}{12} = \underline{\hspace{1cm}}$ $2\frac{5}{9} = \underline{\hspace{1cm}}$ $2\frac{7}{8} = \underline{\hspace{1cm}}$ $1\frac{3}{7} = \underline{\hspace{1cm}}$

$5\frac{2}{5} = \underline{\hspace{1cm}}$ $4\frac{3}{10} = \underline{\hspace{1cm}}$ $1\frac{3}{16} = \underline{\hspace{1cm}}$ $3\frac{2}{9} = \underline{\hspace{1cm}}$

$1\frac{1}{14} = \underline{\hspace{1cm}}$ $5\frac{1}{8} = \underline{\hspace{1cm}}$ $2\frac{2}{15} = \underline{\hspace{1cm}}$ $1\frac{11}{12} = \underline{\hspace{1cm}}$

$4\frac{2}{7} = \underline{\hspace{1cm}}$ $6\frac{7}{10} = \underline{\hspace{1cm}}$ $4\frac{7}{9} = \underline{\hspace{1cm}}$ $3\frac{5}{6} = \underline{\hspace{1cm}}$

FRACTIONS

Write each fraction as a whole or mixed number in simplest form.

EXAMPLE:

$$\frac{8}{5} = 8 \div 5 = 1\frac{3}{5}$$

$\frac{18}{6} =$ _____

$\frac{16}{8} =$ _____

$\frac{12}{5} =$ _____

$\frac{5}{2} =$ _____

$\frac{5}{3} =$ _____

$\frac{13}{8} =$ _____

$\frac{11}{4} =$ _____

$\frac{9}{2} =$ _____

$\frac{8}{3} =$ _____

$\frac{13}{6} =$ _____

$\frac{22}{7} =$ _____

$\frac{18}{4} =$ _____

$\frac{5}{4} =$ _____

$\frac{3}{3} =$ _____

$\frac{14}{8} =$ _____

$\frac{9}{6} =$ _____

$\frac{10}{7} =$ _____

$\frac{6}{4} =$ _____

$\frac{9}{5} =$ _____

$\frac{13}{4} =$ _____

$\frac{8}{5} =$ _____

$\frac{9}{7} =$ _____

$\frac{7}{2} =$ _____

$\frac{15}{2} =$ _____

$\frac{12}{7} =$ _____

$\frac{9}{4} =$ _____

$\frac{9}{8} =$ _____

$\frac{10}{6} =$ _____

$\frac{19}{4} =$ _____

$\frac{7}{6} =$ _____

$\frac{10}{8} =$ _____

$\frac{10}{3} =$ _____

$\frac{8}{4} =$ _____

$\frac{6}{5} =$ _____

$\frac{8}{6} =$ _____

$\frac{4}{4} =$ _____

$\frac{16}{5} =$ _____

$\frac{11}{6} =$ _____

$\frac{7}{5} =$ _____

$\frac{8}{2} =$ _____

$\frac{7}{3} =$ _____

$\frac{11}{7} =$ _____

$\frac{11}{5} =$ _____

$\frac{14}{5} =$ _____

ANSWER KEY FOR MILLIKEN'S FRACTIONS WORKBOOK
FOR
GRADES 3 AND 4

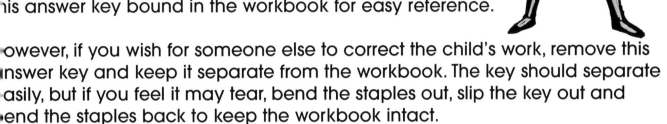

you wish for the child to correct his or her own work, leave
his answer key bound in the workbook for easy reference.

owever, if you wish for someone else to correct the child's work, remove this
nswer key and keep it separate from the workbook. The key should separate
asily, but if you feel it may tear, bend the staples out, slip the key out and
end the staples back to keep the workbook intact.

nswers are set up in order of page number, by rows and columns,
eading left to right.

AGE 1:

$\frac{4}{5}$ $\frac{5}{8}$ $\frac{4}{8}$ (or $\frac{1}{2}$)

$\frac{3}{5}$ $\frac{3}{6}$ (or $\frac{1}{2}$)

AGE 2:

$\frac{2}{5}$

$\frac{5}{8}$ $\frac{3}{7}$

$\frac{4}{9}$ $\frac{3}{10}$

$\frac{5}{7}$ $\frac{5}{6}$

AGE 3:

$\frac{6}{10}$ $\frac{6}{8}$ $\frac{4}{7}$

$\frac{4}{9}$ $\frac{5}{8}$

$\frac{2}{6}$ $\frac{9}{12}$ $\frac{4}{8}$

$\frac{7}{10}$ $\frac{4}{7}$

AGE 4:

MISSILETOE

PAGE 5:

6   3   8   10   12
1   2   4   3   5

$\frac{1}{2}$ $\frac{1}{3}$ $\frac{3}{6}$

$\frac{1}{4}$ $\frac{2}{4}$ $\frac{3}{4}$

$\frac{4}{10}$ $\frac{4}{9}$ $\frac{6}{6}$ or 1

PAGE 6:

$\frac{1}{2} < \frac{2}{3}$, $\frac{4}{5} > \frac{1}{2}$

$\frac{2}{4} = \frac{1}{2}$

$\frac{3}{4} > \frac{1}{2}$, $\frac{6}{8} = \frac{3}{4}$,

$\frac{2}{3} < \frac{5}{6}$

$\frac{3}{6} < \frac{4}{5}$, $\frac{1}{2} < \frac{3}{4}$,

$\frac{4}{6} = \frac{2}{3}$

PAGE 7:

$\frac{1}{2}$ $\frac{1}{3}$ $\frac{1}{2}$

$\frac{2}{3}$ $\frac{3}{4}$ $\frac{2}{5}$

$\frac{2}{4} = \frac{1}{2}$ $\frac{2}{6} = \frac{1}{3}$

PAGE 8:

Children should color the
regions according to the
directions on the page.

PAGE 9:

Children should match
equivalent fractions.

$\frac{1}{2} = \frac{2}{4}$; $\frac{1}{3} = \frac{2}{6}$

$\frac{2}{3} = \frac{4}{6}$; $\frac{3}{4} = \frac{6}{8}$

$\frac{4}{5} = \frac{8}{10}$; $\frac{1}{4} = \frac{2}{8}$

$\frac{1}{5} = \frac{2}{10}$; $\frac{4}{6} = \frac{8}{12}$

PAGE 10:

$\frac{2}{8}$ or $\frac{1}{4}$, $\frac{4}{6}$ or $\frac{2}{3}$,

$\frac{2}{8}$ or $\frac{1}{4}$, $\frac{3}{6}$ or $\frac{1}{2}$;

$\frac{4}{8}$ or $\frac{1}{2}$, $\frac{2}{6}$ or $\frac{1}{3}$,

$\frac{2}{4}$ or $\frac{1}{2}$, $\frac{2}{6}$ or $\frac{1}{3}$

$\frac{1}{2}$ is used most often.

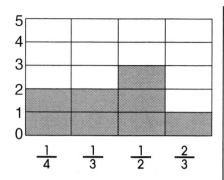

PAGE 11:
Children should color the regions according to the directions on the page.

PAGE 12:
$$\frac{1}{2}$$

$$\frac{1}{3} \quad \frac{2}{3}$$

$$\frac{1}{4} \quad \frac{2}{4} \quad \frac{3}{4}$$

PAGE 13:
$3\frac{1}{2}$; $2\frac{1}{2}$; $\frac{1}{2}$; 3 ; 2

4 ; 2 ; 12 ; 6 ; 10

PAGE 14:
Children should color the regions according to the directions on the page.

PAGE 15:
$$\frac{4}{6} \quad \frac{2}{10}$$

$$\frac{4}{10} \quad \frac{2}{4}$$

$$\frac{6}{12} \quad \frac{2}{6} \quad \frac{9}{12} \quad \frac{6}{12}$$

$$\frac{1}{3} \quad \frac{2}{6} \quad \frac{1}{3} \quad \frac{3}{9}$$

$$\frac{6}{18} \quad \frac{2}{6} \quad \frac{5}{15} \quad \frac{4}{12}$$

PAGE 16:
$$\frac{2}{6} \quad \frac{8}{10} \quad \frac{6}{20} \quad \frac{2}{12}$$

$$\frac{21}{30} \quad \frac{6}{9} \quad \frac{3}{12} \quad \frac{3}{15}$$

$$\frac{12}{32} \quad \frac{36}{40} \quad \frac{4}{12} \quad \frac{28}{32}$$

$$\frac{3}{5} \quad \frac{1}{4}$$

PAGE 16: (continued)
$$\frac{2}{3} \quad \frac{7}{8} \quad \frac{3}{10}$$

$$\frac{3}{4} \quad \frac{4}{5} \quad \frac{1}{3} \quad \frac{5}{6}$$

$$\frac{1}{5} \quad \frac{1}{3} \quad \frac{1}{6} \quad \frac{3}{20}$$

$$\frac{1}{6} \quad \frac{2}{5} \quad \frac{1}{9} \quad \frac{1}{3}$$

PAGE 17:
Children should color the regions according to the directions on the page.

PAGE 18:
$$\frac{3}{4} \quad \frac{1}{2} \quad \frac{1}{4}$$

$$\frac{1}{5} \quad \frac{2}{3} \quad \frac{1}{2}$$

$$\frac{2}{3} \quad \frac{1}{3} \quad \frac{2}{3}$$

$$\frac{8}{9} \quad \frac{1}{2} \quad \frac{5}{7} \quad \frac{1}{6}$$

$$\frac{7}{8} \quad \frac{3}{7} \quad \frac{2}{7} \quad \frac{5}{7}$$

$$\frac{1}{3} \quad \frac{5}{6} \quad \frac{2}{11} \quad \frac{2}{3}$$

$$\frac{5}{18} \quad \frac{3}{8} \quad \frac{5}{6} \quad \frac{8}{9}$$

$$\frac{2}{9} \quad \frac{6}{11} \quad \frac{2}{3} \quad \frac{2}{5}$$

$$\frac{3}{4} \quad \frac{5}{8} \quad \frac{1}{3} \quad \frac{7}{8}$$

$$\frac{4}{15} \quad \frac{1}{3} \quad \frac{7}{10} \quad \frac{5}{6}$$

$$\frac{7}{12} \quad \frac{11}{18} \quad \frac{5}{6} \quad \frac{5}{8}$$

PAGE 19:
2   3
3   2
4   4   9   15
6   30   32   27
20   24   21   36
16   33   40   24
2, 3, 4, 5, 6; 18, 27, 36, 45, 54
4, 6, 8, 10, 12; 24, 36, 48, 60, 72
2, 3, 4, 5, 6; 20, 30, 40, 50, 60
10, 15, 20, 25, 30; 14, 12, 16, 35, 24
2, 3, 4, 5, 6; 10, 24, 20, 40, 30

PAGE 20:
Children should color the regions according to the directions on the page.

PAGE 21:
$$\frac{12}{5} \quad \frac{25}{8}$$

$$\frac{7}{6} \quad \frac{17}{4} \quad \frac{16}{5}$$

$$\frac{8}{3} \quad \frac{53}{9} \quad \frac{37}{8}$$

$$\frac{16}{3} \quad \frac{9}{2} \quad \frac{22}{7}$$

$$\frac{15}{4} \quad \frac{17}{6} \quad \frac{14}{5}$$

$$\frac{21}{10} \quad \frac{13}{9} \quad \frac{25}{16}$$

$$\frac{19}{12} \quad \frac{18}{7} \quad \frac{11}{8}$$

$$\frac{55}{9} \quad \frac{23}{5} \quad \frac{39}{10}$$

$$\frac{29}{12} \quad \frac{23}{9} \quad \frac{23}{8} \quad \frac{10}{7}$$

$$\frac{27}{5} \quad \frac{43}{10} \quad \frac{19}{16} \quad \frac{29}{9}$$

$$\frac{15}{14} \quad \frac{41}{8} \quad \frac{32}{15} \quad \frac{23}{12}$$

$$\frac{30}{7} \quad \frac{67}{10} \quad \frac{43}{9} \quad \frac{23}{6}$$

PAGE 22:
$$1\frac{3}{5} \quad 2$$

$$1\frac{2}{7} \quad 1\frac{1}{5}$$

$$3 \quad 3\frac{1}{7} \quad 3\frac{1}{2} \quad 1\frac{1}{3}$$

$$2 \quad 4\frac{1}{2} \quad 7\frac{1}{2} \quad 1$$

$$2\frac{2}{5} \quad 1\frac{1}{4} \quad 1\frac{5}{7} \quad 3\frac{1}{5}$$

$$2\frac{1}{2} \quad 1 \quad 2\frac{1}{4} \quad 1\frac{5}{6}$$

$$1\frac{2}{3} \quad 1\frac{3}{4} \quad 1\frac{1}{8} \quad 1\frac{2}{5}$$

$$1\frac{5}{8} \quad 1\frac{1}{2} \quad 1\frac{2}{3} \quad 4$$

$$2\frac{3}{4} \quad 1\frac{3}{7} \quad 4\frac{3}{4} \quad 2\frac{1}{3}$$

$$4\frac{1}{2} \quad 1\frac{1}{2} \quad 1\frac{1}{6} \quad 1\frac{4}{7}$$

$$2\frac{2}{3} \quad 1\frac{4}{5} \quad 1\frac{1}{4} \quad 2\frac{1}{5}$$

$$2\frac{1}{6} \quad 3\frac{1}{4} \quad 3\frac{1}{3} \quad 2\frac{4}{5}$$

GE 23:

2, 6, 12, 3
2, 12, 4, 3
4, 8, 2, 1
10, 6, 14, 3

$1\frac{2}{3}$ 2 $1\frac{1}{4}$ $1\frac{1}{3}$

$2\frac{1}{4}$ $4\frac{1}{2}$ 1 $2\frac{1}{6}$

$1\frac{1}{2}$ $1\frac{1}{3}$ $1\frac{3}{4}$ $2\frac{1}{2}$

$3\frac{1}{3}$ $3\frac{1}{2}$ $1\frac{1}{8}$ $1\frac{4}{5}$

$\frac{9}{8}$ $\frac{9}{4}$ $\frac{11}{3}$ $\frac{11}{6}$

$\frac{11}{5}$ $\frac{23}{6}$ $\frac{21}{16}$ $\frac{17}{10}$

$\frac{7}{4}$ $\frac{7}{3}$ $\frac{3}{2}$ $\frac{7}{2}$

$\frac{19}{6}$ $\frac{15}{8}$ $\frac{19}{8}$ $\frac{13}{8}$

GE 24:

Children should color the regions according to the directions on the page.

GE 25:

= > =
< = <
> < >
= < = >
> < = =
< = > <
< > < =
= < < =
> = < <
= = < >

GE 26:

$\frac{3}{4}$, $\frac{1}{2}$

$\frac{1}{3}$, $\frac{4}{5}$

$\frac{7}{9}$, $\frac{9}{11}$, $1\frac{1}{5}$, $\frac{1}{2}$

$1\frac{3}{7}$, 1, $\frac{4}{5}$, 1

$\frac{1}{2}$, $\frac{9}{11}$, $1\frac{2}{5}$, $1\frac{1}{3}$

1, $\frac{11}{15}$, $1\frac{1}{7}$, $\frac{4}{7}$

$\frac{4}{7}$, 1, $\frac{3}{4}$, 1

PAGE 26: (continued)

1, $1\frac{3}{5}$, $\frac{3}{5}$, $1\frac{1}{3}$

$1\frac{4}{5}$, 1, $\frac{5}{8}$, $\frac{3}{7}$

$\frac{3}{5}$, $1\frac{2}{5}$, 1, $\frac{6}{7}$

$\frac{1}{9}$, $1\frac{2}{9}$, $\frac{1}{3}$, $1\frac{1}{4}$

PAGE 27:

$\frac{2}{3}$, $\frac{2}{3}$, $1\frac{1}{5}$, $\frac{1}{4}$,
$\frac{1}{2}$, $\frac{2}{5}$

$\frac{2}{3}$, $\frac{3}{4}$, $1\frac{2}{7}$, 1, $\frac{2}{3}$

1, $\frac{4}{5}$, $\frac{8}{15}$, $1\frac{1}{3}$, $\frac{5}{7}$,

$\frac{2}{7}$, $\frac{4}{5}$, $\frac{1}{2}$, $\frac{10}{11}$, $1\frac{3}{4}$,
$\frac{8}{9}$

$\frac{1}{2}$, $\frac{5}{7}$, 1, $\frac{2}{5}$, $1\frac{1}{10}$,
$\frac{2}{3}$

$\frac{5}{6}$, $1\frac{3}{10}$, $\frac{1}{2}$, $\frac{5}{7}$, $\frac{3}{5}$,
$\frac{5}{7}$

PAGE 28:

1, $1\frac{1}{4}$

$1\frac{1}{5}$, $1\frac{1}{3}$

1, $1\frac{1}{6}$, 1, $1\frac{1}{3}$

$1\frac{1}{8}$, $1\frac{1}{2}$, $1\frac{3}{8}$, $1\frac{3}{5}$

$1\frac{1}{4}$, $1\frac{5}{8}$, $1\frac{1}{2}$, 1

$1\frac{1}{3}$, $1\frac{1}{2}$, $1\frac{1}{4}$, $1\frac{1}{3}$

$\frac{6}{7}$, $\frac{3}{4}$, $1\frac{3}{8}$, $\frac{2}{3}$, $1\frac{2}{7}$,
$1\frac{1}{7}$

1, $1\frac{2}{3}$, $1\frac{2}{3}$, $1\frac{1}{2}$, $\frac{9}{10}$,
$1\frac{1}{16}$

$1\frac{1}{3}$, $1\frac{1}{4}$, $\frac{3}{5}$, $1\frac{1}{5}$,
$\frac{7}{8}$, $\frac{3}{4}$

PAGE 29:

$\frac{1}{3}$	$\frac{2}{3}$	$\frac{2}{5}$	$\frac{4}{9}$
$\frac{3}{7}$	$\frac{1}{6}$	$\frac{3}{11}$	$\frac{2}{5}$
$\frac{4}{7}$	$\frac{1}{5}$	$\frac{3}{4}$	$\frac{3}{7}$
$\frac{2}{9}$	$\frac{1}{10}$	$\frac{5}{7}$	$\frac{2}{5}$
$\frac{5}{11}$	$\frac{1}{3}$	$\frac{3}{5}$	$\frac{1}{2}$
$\frac{2}{7}$	$\frac{1}{3}$	$\frac{3}{8}$	$\frac{1}{7}$
$\frac{3}{7}$	$\frac{1}{3}$	$\frac{4}{11}$	$\frac{1}{5}$
$\frac{1}{4}$	$\frac{4}{7}$	$\frac{1}{3}$	$\frac{1}{2}$
$\frac{1}{3}$	$\frac{2}{9}$	$\frac{1}{3}$	$\frac{1}{11}$
$\frac{1}{5}$	$\frac{2}{3}$	$\frac{1}{10}$	$\frac{1}{7}$

PAGE 30:

$\frac{1}{2}$ $\frac{1}{5}$

$\frac{1}{2}$ $\frac{1}{3}$

$\frac{1}{4}$ $\frac{1}{2}$ $\frac{1}{3}$ $\frac{5}{8}$

$\frac{2}{5}$ $\frac{1}{8}$ $\frac{1}{8}$ $\frac{2}{5}$

$\frac{2}{5}$ $\frac{1}{4}$ $\frac{2}{3}$ $\frac{1}{5}$ $\frac{2}{7}$ $\frac{5}{8}$

$\frac{1}{2}$ $\frac{1}{4}$ $\frac{1}{5}$ $\frac{3}{4}$ $\frac{2}{5}$ $\frac{1}{4}$

$\frac{3}{10}$ $\frac{3}{5}$ $\frac{1}{7}$ $\frac{3}{10}$ $\frac{3}{16}$ $\frac{1}{4}$

$\frac{1}{4}$ $\frac{1}{10}$ $\frac{1}{7}$ $\frac{1}{2}$ $\frac{2}{9}$ $\frac{1}{2}$

PAGE 31:

$\frac{1}{4}$ $\frac{4}{5}$ $\frac{1}{2}$ $\frac{1}{9}$ $\frac{1}{5}$ $\frac{1}{3}$

$\frac{1}{5}$ $\frac{1}{9}$ $\frac{3}{7}$ $\frac{1}{7}$ $\frac{2}{11}$

$\frac{2}{7}$ $\frac{1}{4}$ $\frac{1}{4}$ $\frac{4}{5}$ $\frac{2}{5}$

$\frac{1}{11}$ $\frac{3}{10}$ $\frac{1}{2}$ $\frac{3}{7}$ $\frac{3}{8}$ $\frac{5}{9}$

$\frac{8}{21}$ $\frac{5}{6}$ $\frac{4}{7}$ $\frac{1}{2}$ $\frac{5}{11}$ $\frac{2}{11}$

$\frac{1}{5}$ $\frac{1}{15}$ $\frac{1}{2}$ $\frac{5}{11}$ $\frac{1}{4}$ $\frac{4}{13}$

PAGE 32:

1. $\frac{2}{3}$ 5. $\frac{5}{8}$

2. $\frac{2}{5}$ 6. $\frac{7}{9}$

3. $\frac{2}{3}$ 7. $\frac{1}{5}$

4. $\frac{1}{2}$

PAGE 33:

8. $\frac{3}{8}$ 10. $\frac{5}{8}$

9. $\frac{1}{5}$ 11. $\frac{7}{10}$

HE HAD NO BODY !

PAGE 34:

$8\frac{2}{3}$, $14\frac{4}{7}$

$13\frac{1}{2}$, $9\frac{1}{3}$, $6\frac{1}{3}$, 4, 9

$7\frac{1}{2}$, 4, $7\frac{1}{2}$, $6\frac{2}{3}$, 10

$4\frac{1}{4}$, $3\frac{4}{5}$, 6, $9\frac{2}{5}$, 8

12, 6, $10\frac{3}{4}$, 6, $15\frac{3}{4}$

PAGE 35:

$\frac{2}{5}$, $\frac{2}{3}$

$\frac{7}{10}$, $1\frac{2}{3}$

$\frac{2}{3}$, $\frac{1}{3}$

$\frac{5}{8}$, $\frac{3}{5}$, $1\frac{1}{8}$, 1

$\frac{3}{4}$, $\frac{5}{8}$, $\frac{1}{2}$, $1\frac{1}{4}$

$26\frac{5}{6}$, $34\frac{3}{7}$, $23\frac{5}{6}$,

$6\frac{1}{3}$, $13\frac{1}{4}$

$15\frac{1}{8}$, $19\frac{1}{3}$, $8\frac{2}{3}$,

$17\frac{1}{2}$, $9\frac{3}{5}$

$14\frac{5}{8}$, 16, 12, 10, 17

PAGE 36:

1. $\frac{3}{10}$ 6. $\frac{1}{4}$

2. $\frac{8}{9}$ 7. $\frac{2}{5}$

3. $\frac{1}{2}$ 8. $\frac{1}{3}$

4. $\frac{3}{7}$ 9. $\frac{5}{7}$

5. $\frac{1}{3}$ 10. $\frac{1}{4}$

PAGE 37:

36
30,   48, 140
72, 135,   45
22,   72,   44
75, 140, 210

PAGE 38:

6
5,   8,   4
27, 13,   3
6,   7,   3
16,   6,   9

PAGE 39:

$1\frac{1}{10}$, $\frac{5}{6}$

$1\frac{1}{4}$, $1\frac{1}{8}$, $1\frac{8}{21}$, $1\frac{1}{12}$

$\frac{11}{18}$, $\frac{7}{12}$, $\frac{13}{14}$, $\frac{14}{15}$

$1\frac{3}{10}$, $1\frac{2}{9}$, $\frac{17}{24}$, $1\frac{7}{12}$

$1\frac{1}{2}$, $1\frac{5}{18}$, $1\frac{5}{9}$, $1\frac{7}{16}$

PAGE 40:

$1\frac{1}{4}$, $\frac{19}{24}$

$\frac{7}{8}$, $1\frac{1}{6}$, $\frac{9}{10}$, $\frac{7}{9}$

$1\frac{1}{6}$, $1\frac{4}{15}$, $1\frac{7}{36}$, $\frac{8}{9}$

$\frac{11}{20}$, $1\frac{1}{16}$, $1\frac{7}{20}$, $\frac{15}{16}$

$\frac{35}{36}$, $1\frac{1}{3}$, $1\frac{7}{16}$, $1\frac{1}{8}$

PAGE 41:

$1\frac{1}{5}$, $1\frac{1}{7}$, $1\frac{1}{12}$, $1\frac{1}{9}$

$1\frac{3}{8}$, $1\frac{3}{4}$, 1, 2, $\frac{1}{2}$, $1\frac{1}{2}$

$1\frac{1}{24}$, $1\frac{11}{18}$, $1\frac{1}{6}$, $1\frac{7}{9}$,

$1\frac{14}{15}$

$1\frac{5}{42}$, 2, $\frac{9}{10}$, $1\frac{1}{9}$, $2\frac{3}{16}$

PAGE 42:

$\frac{1}{2}$

$\frac{3}{10}$ $\frac{1}{6}$ $\frac{1}{8}$

$\frac{3}{14}$ $\frac{7}{24}$ $\frac{1}{10}$

$\frac{2}{5}$ $\frac{7}{12}$ $\frac{9}{40}$

$\frac{11}{24}$ $\frac{1}{5}$ $\frac{5}{12}$

$\frac{4}{9}$ $\frac{8}{15}$ $\frac{3}{20}$

PAGE 43:

$\frac{1}{3}$ $\frac{1}{6}$

$\frac{9}{14}$ $\frac{2}{9}$

$\frac{1}{3}$ $\frac{9}{20}$ $\frac{2}{15}$

$\frac{7}{15}$ $\frac{1}{4}$ $\frac{5}{18}$

$\frac{1}{5}$ $\frac{8}{21}$ $\frac{1}{12}$

$\frac{1}{4}$ $\frac{1}{2}$ $\frac{1}{4}$

PAGE 44:

$\frac{19}{30}$, $\frac{7}{8}$, $\frac{1}{2}$, $1\frac{1}{10}$, $1\frac{1}{6}$

$\frac{8}{9}$, $1\frac{5}{12}$, $\frac{9}{10}$, 1, $\frac{1}{2}$

$\frac{1}{10}$, $\frac{1}{4}$, $\frac{2}{5}$, $\frac{1}{6}$, $\frac{11}{40}$

$\frac{1}{6}$, $\frac{1}{6}$, $\frac{3}{10}$, $\frac{13}{24}$, $\frac{1}{3}$

$\frac{5}{7}$, $\frac{2}{9}$, $\frac{11}{20}$, $\frac{1}{5}$, $\frac{5}{12}$

EQUIVALENT FRACTIONS.

Write the correct number in each blank.

$$\frac{}{6} = \frac{1}{3} \qquad \frac{3}{4} = \frac{}{8} \qquad \frac{3}{4} = \frac{}{16} \qquad \frac{}{2} = \frac{9}{6}$$

$$\frac{1}{2} = \frac{}{4} \qquad \frac{6}{5} = \frac{}{10} \qquad \frac{}{8} = \frac{1}{2} \qquad \frac{12}{16} = \frac{}{4}$$

$$\frac{2}{3} = \frac{}{6} \qquad \frac{4}{4} = \frac{}{8} \qquad \frac{1}{4} = \frac{}{8} \qquad \frac{3}{6} = \frac{}{2}$$

$$\frac{5}{8} = \frac{}{16} \qquad \frac{3}{2} = \frac{}{4} \qquad \frac{7}{8} = \frac{}{16} \qquad \frac{9}{24} = \frac{}{8}$$

EXAMPLE:

$$\frac{2}{6} = \frac{2 \div 2}{6 \div 2} = \frac{1}{3}$$

OR:

$$1\frac{1}{8} = 1 + \frac{1}{8} =$$

$$\frac{8}{8} + \frac{1}{8} = \frac{9}{8}$$

Write each fraction as a whole or mixed number in simplest form.

$$\frac{5}{3} = \underline{\hspace{1cm}} \qquad \frac{8}{4} = \underline{\hspace{1cm}} \qquad \frac{10}{8} = \underline{\hspace{1cm}} \qquad \frac{8}{6} = \underline{\hspace{1cm}}$$

$$\frac{9}{4} = \underline{\hspace{1cm}} \qquad \frac{9}{2} = \underline{\hspace{1cm}} \qquad \frac{4}{4} = \underline{\hspace{1cm}} \qquad \frac{13}{6} = \underline{\hspace{1cm}}$$

$$\frac{3}{2} = \underline{\hspace{1cm}} \qquad \frac{4}{3} = \underline{\hspace{1cm}} \qquad \frac{7}{4} = \underline{\hspace{1cm}} \qquad \frac{5}{2} = \underline{\hspace{1cm}}$$

$$\frac{10}{3} = \underline{\hspace{1cm}} \qquad \frac{7}{2} = \underline{\hspace{1cm}} \qquad \frac{9}{8} = \underline{\hspace{1cm}} \qquad \frac{9}{5} = \underline{\hspace{1cm}}$$

Write each mixed numeral as a fraction.

$$1\frac{1}{8} = \underline{\hspace{1cm}} \qquad 2\frac{1}{4} = \underline{\hspace{1cm}} \qquad 3\frac{2}{3} = \underline{\hspace{1cm}} \qquad 1\frac{5}{6} = \underline{\hspace{1cm}}$$

$$2\frac{1}{5} = \underline{\hspace{1cm}} \qquad 3\frac{5}{6} = \underline{\hspace{1cm}} \qquad 1\frac{5}{16} = \underline{\hspace{1cm}} \qquad 1\frac{7}{10} = \underline{\hspace{1cm}}$$

$$1\frac{3}{4} = \underline{\hspace{1cm}} \qquad 2\frac{1}{3} = \underline{\hspace{1cm}} \qquad 1\frac{1}{2} = \underline{\hspace{1cm}} \qquad 3\frac{1}{2} = \underline{\hspace{1cm}}$$

$$3\frac{1}{6} = \underline{\hspace{1cm}} \qquad 1\frac{7}{8} = \underline{\hspace{1cm}} \qquad 2\frac{3}{8} = \underline{\hspace{1cm}} \qquad 1\frac{5}{8} = \underline{\hspace{1cm}}$$

23

Shade in all mixed numerals in their simplest form.

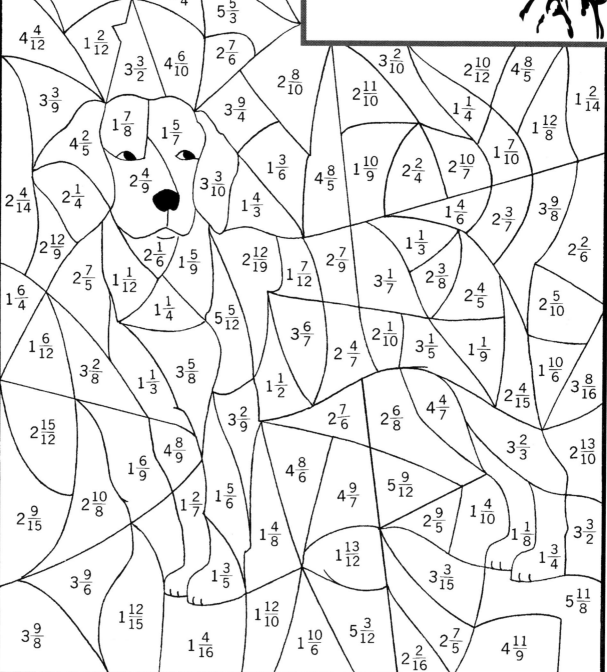

COMPARING FRACTIONS.

Write **<**, **>**, or **=** between the fractions to make each sentence true.

$\dfrac{5}{6}$	$\dfrac{15}{18}$	$\dfrac{3}{4}$	$\dfrac{1}{2}$	$\dfrac{4}{7}$	$\dfrac{36}{63}$		
$\dfrac{9}{10}$	$\dfrac{11}{12}$	$\dfrac{8}{15}$	$\dfrac{16}{30}$	$\dfrac{3}{4}$	$\dfrac{6}{7}$		
$\dfrac{9}{12}$	$\dfrac{6}{13}$	$\dfrac{7}{8}$	$\dfrac{10}{11}$	$\dfrac{5}{6}$	$\dfrac{4}{5}$		
$\dfrac{22}{24}$	$\dfrac{11}{12}$	$\dfrac{3}{5}$	$\dfrac{5}{8}$	$\dfrac{24}{60}$	$\dfrac{2}{5}$	$\dfrac{2}{5}$	$\dfrac{1}{4}$
$\dfrac{4}{5}$	$\dfrac{2}{3}$	$\dfrac{1}{6}$	$\dfrac{5}{24}$	$\dfrac{4}{5}$	$\dfrac{32}{40}$	$\dfrac{9}{10}$	$\dfrac{36}{40}$
$\dfrac{4}{9}$	$\dfrac{7}{14}$	$\dfrac{5}{9}$	$\dfrac{25}{45}$	$\dfrac{5}{8}$	$\dfrac{4}{7}$	$\dfrac{1}{3}$	$\dfrac{2}{5}$
$\dfrac{10}{75}$	$\dfrac{2}{5}$	$\dfrac{6}{10}$	$\dfrac{5}{12}$	$\dfrac{5}{9}$	$\dfrac{10}{16}$	$\dfrac{9}{12}$	$\dfrac{27}{36}$
$\dfrac{9}{60}$	$\dfrac{3}{20}$	$\dfrac{7}{12}$	$\dfrac{7}{11}$	$\dfrac{7}{15}$	$\dfrac{8}{15}$	$\dfrac{2}{15}$	$\dfrac{10}{75}$
$\dfrac{8}{9}$	$\dfrac{21}{24}$	$\dfrac{33}{55}$	$\dfrac{3}{5}$	$\dfrac{6}{8}$	$\dfrac{20}{25}$	$\dfrac{3}{11}$	$\dfrac{4}{10}$
$\dfrac{3}{10}$	$\dfrac{21}{70}$	$\dfrac{25}{35}$	$\dfrac{5}{7}$	$\dfrac{8}{16}$	$\dfrac{8}{12}$	$\dfrac{7}{12}$	$\dfrac{5}{9}$

25

EXAMPLE:

$$\frac{5}{8} + \frac{1}{8} = \frac{5+1}{8} = \frac{6}{8} = \frac{3}{4}$$

Write each sum in its simplest form.

$\frac{5}{8} + \frac{1}{8} =$ _____ $\frac{1}{4} + \frac{1}{4} =$ _____

$\frac{1}{6} + \frac{1}{6} =$ _____ $\frac{3}{5} + \frac{1}{5} =$ _____

$\frac{2}{9} + \frac{5}{9} =$ _____ $\frac{5}{11} + \frac{4}{11} =$ _____ $\frac{3}{10} + \frac{9}{10} =$ _____ $\frac{5}{12} + \frac{1}{12} =$ _____

$\frac{4}{7} + \frac{6}{7} =$ _____ $\frac{5}{6} + \frac{1}{6} =$ _____ $\frac{2}{5} + \frac{2}{5} =$ _____ $\frac{1}{10} + \frac{9}{10} =$ _____

$\frac{3}{8} + \frac{1}{8} =$ _____ $\frac{7}{11} + \frac{2}{11} =$ _____ $\frac{4}{5} + \frac{3}{5} =$ _____ $\frac{4}{9} + \frac{8}{9} =$ _____

$\frac{7}{12} + \frac{5}{12} =$ _____ $\frac{4}{15} + \frac{7}{15} =$ _____ $\frac{2}{7} + \frac{6}{7} =$ _____ $\frac{1}{7} + \frac{3}{7} =$ _____

$\frac{5}{14} + \frac{3}{14} =$ _____ $\frac{7}{16} + \frac{9}{16} =$ _____ $\frac{3}{8} + \frac{3}{8} =$ _____ $\frac{7}{9} + \frac{2}{9} =$ _____

$\frac{5}{11} + \frac{6}{11} =$ _____ $\frac{7}{10} + \frac{9}{10} =$ _____ $\frac{8}{15} + \frac{1}{15} =$ _____ $\frac{11}{12} + \frac{5}{12} =$ _____

$\frac{9}{10} + \frac{9}{10} =$ _____ $\frac{3}{7} + \frac{4}{7} =$ _____ $\frac{5}{16} + \frac{5}{16} =$ _____ $\frac{1}{14} + \frac{5}{14} =$ _____

$\frac{7}{15} + \frac{2}{15} =$ _____ $\frac{7}{10} + \frac{7}{10} =$ _____ $\frac{4}{11} + \frac{7}{11} =$ _____ $\frac{9}{14} + \frac{3}{14} =$ _____

$\frac{1}{18} + \frac{1}{18} =$ _____ $\frac{7}{9} + \frac{4}{9} =$ _____ $\frac{5}{18} + \frac{1}{18} =$ _____ $\frac{3}{8} + \frac{7}{8} =$ _____

26

$\frac{4}{9}$ $+\frac{2}{9}$	$\frac{1}{3}$ $+\frac{1}{3}$	$\frac{4}{5}$ $+\frac{2}{5}$	$\frac{1}{8}$ $+\frac{1}{8}$	$\frac{3}{20}$ $+\frac{7}{20}$	$\frac{2}{15}$ $+\frac{4}{15}$
$\frac{7}{12}$ $+\frac{1}{12}$	$\frac{9}{16}$ $+\frac{3}{16}$	$\frac{3}{7}$ $+\frac{6}{7}$	$\frac{9}{11}$ $+\frac{2}{11}$	$\frac{5}{18}$ $+\frac{7}{18}$	
$\frac{1}{12}$ $+\frac{11}{12}$	$\frac{1}{10}$ $+\frac{7}{10}$	$\frac{7}{15}$ $+\frac{1}{15}$	$\frac{7}{9}$ $+\frac{5}{9}$	$\frac{9}{14}$ $+\frac{1}{14}$	
$\frac{4}{21}$ $+\frac{2}{21}$	$\frac{9}{20}$ $+\frac{7}{20}$	$\frac{7}{16}$ $+\frac{1}{16}$	$\frac{3}{11}$ $+\frac{7}{11}$	$\frac{7}{8}$ $+\frac{7}{8}$	$\frac{5}{18}$ $+\frac{11}{18}$
$\frac{7}{24}$ $+\frac{5}{24}$	$\frac{5}{21}$ $+\frac{10}{21}$	$\frac{11}{14}$ $+\frac{3}{14}$	$\frac{8}{25}$ $+\frac{2}{25}$	$\frac{9}{20}$ $+\frac{13}{20}$	$\frac{2}{15}$ $+\frac{8}{15}$
$\frac{5}{12}$ $+\frac{5}{12}$	$\frac{17}{20}$ $+\frac{9}{20}$	$\frac{11}{24}$ $+\frac{1}{24}$	$\frac{5}{21}$ $+\frac{10}{21}$	$\frac{7}{30}$ $+\frac{11}{30}$	$\frac{13}{21}$ $+\frac{2}{21}$

Write each sum in its simplest form.

EXAMPLE:

$\frac{4}{9}$
$+\frac{2}{9}$

$\frac{6}{9} = \frac{2}{3}$

27

$\dfrac{7}{8} + \dfrac{1}{8} =$ ___ $\dfrac{3}{4} + \dfrac{2}{4} =$ ___

Write each sum in its simplest form.

EXAMPLE:

$$\dfrac{7}{8} + \dfrac{1}{8} = \dfrac{7+1}{8} = \dfrac{8}{8} = 1$$

$\dfrac{2}{5} + \dfrac{4}{5} =$ ___ $\dfrac{3}{6} + \dfrac{5}{6} =$ ___

$\dfrac{2}{4} + \dfrac{2}{4} =$ ___ $\dfrac{3}{6} + \dfrac{4}{6} =$ ___ $\dfrac{1}{2} + \dfrac{1}{2} =$ ___ $\dfrac{2}{3} + \dfrac{2}{3} =$ ___

$\dfrac{4}{8} + \dfrac{5}{8} =$ ___ $\dfrac{3}{4} + \dfrac{3}{4} =$ ___ $\dfrac{5}{8} + \dfrac{6}{8} =$ ___ $\dfrac{4}{5} + \dfrac{4}{5} =$ ___

$\dfrac{3}{4} + \dfrac{2}{4} =$ ___ $\dfrac{7}{8} + \dfrac{6}{8} =$ ___ $\dfrac{4}{6} + \dfrac{5}{6} =$ ___ $\dfrac{3}{8} + \dfrac{5}{8} =$ ___

$\dfrac{2}{3} + \dfrac{2}{3} =$ ___ $\dfrac{5}{8} + \dfrac{7}{8} =$ ___ $\dfrac{7}{12} + \dfrac{8}{12} =$ ___ $\dfrac{7}{9} + \dfrac{5}{9} =$ ___

$\dfrac{2}{7}$	$\dfrac{2}{4}$	$\dfrac{4}{8}$	$\dfrac{2}{6}$	$\dfrac{4}{7}$	$\dfrac{5}{7}$
$+\dfrac{4}{7}$	$+\dfrac{1}{4}$	$+\dfrac{7}{8}$	$+\dfrac{2}{6}$	$+\dfrac{5}{7}$	$+\dfrac{3}{7}$
$\dfrac{6}{7}$	$\dfrac{7}{9}$	$\dfrac{5}{6}$	$\dfrac{3}{4}$	$\dfrac{5}{10}$	$\dfrac{8}{16}$
$+\dfrac{1}{7}$	$+\dfrac{8}{9}$	$+\dfrac{5}{6}$	$+\dfrac{3}{4}$	$+\dfrac{4}{10}$	$+\dfrac{9}{16}$
$\dfrac{4}{6}$	$\dfrac{6}{8}$	$\dfrac{3}{10}$	$\dfrac{3}{5}$	$\dfrac{5}{8}$	$\dfrac{7}{16}$
$+\dfrac{4}{6}$	$+\dfrac{4}{8}$	$+\dfrac{3}{10}$	$+\dfrac{3}{5}$	$+\dfrac{2}{8}$	$+\dfrac{5}{16}$

28

SUBTRACTING FRACTIONS.

Write each difference in its simplest form.

$\dfrac{5}{9} - \dfrac{2}{9} =$ _____ $\dfrac{7}{9} - \dfrac{1}{9} =$ _____ $\dfrac{4}{5} - \dfrac{2}{5} =$ _____ $\dfrac{8}{9} - \dfrac{4}{9} =$ _____

$\dfrac{6}{7} - \dfrac{3}{7} =$ _____ $\dfrac{7}{12} - \dfrac{5}{12} =$ _____ $\dfrac{10}{11} - \dfrac{7}{11} =$ _____ $\dfrac{13}{15} - \dfrac{7}{15} =$ _____

$\dfrac{6}{7} - \dfrac{2}{7} =$ _____ $\dfrac{9}{10} - \dfrac{7}{10} =$ _____ $\dfrac{15}{16} - \dfrac{3}{16} =$ _____ $\dfrac{9}{14} - \dfrac{3}{14} =$ _____

$\dfrac{5}{18} - \dfrac{1}{18} =$ _____ $\dfrac{9}{20} - \dfrac{7}{20} =$ _____ $\dfrac{6}{7} - \dfrac{1}{7} =$ _____ $\dfrac{8}{15} - \dfrac{2}{15} =$ _____

$\dfrac{9}{11} - \dfrac{4}{11} =$ _____ $\dfrac{8}{9} - \dfrac{5}{9} =$ _____ $\dfrac{7}{10} - \dfrac{1}{10} =$ _____ $\dfrac{11}{20} - \dfrac{1}{20} =$ _____

$\dfrac{5}{14} - \dfrac{1}{14} =$ _____ $\dfrac{13}{18} - \dfrac{7}{18} =$ _____ $\dfrac{7}{16} - \dfrac{1}{16} =$ _____ $\dfrac{11}{14} - \dfrac{9}{14} =$ _____

$\dfrac{10}{21} - \dfrac{1}{21} =$ _____ $\dfrac{11}{12} - \dfrac{7}{12} =$ _____

EXAMPLE:

$$\dfrac{5}{9} - \dfrac{2}{9} = \dfrac{5-2}{9} = \dfrac{3}{9} = \dfrac{1}{3}$$

$\dfrac{7}{11} - \dfrac{3}{11} =$ _____ $\dfrac{4}{15} - \dfrac{1}{15} =$ _____

$\dfrac{5}{16} - \dfrac{1}{16} =$ _____ $\dfrac{5}{7} - \dfrac{1}{7} =$ _____ $\dfrac{7}{9} - \dfrac{4}{9} =$ _____ $\dfrac{17}{20} - \dfrac{7}{20} =$ _____

$\dfrac{4}{9} - \dfrac{1}{9} =$ _____ $\dfrac{11}{18} - \dfrac{7}{18} =$ _____ $\dfrac{5}{12} - \dfrac{1}{12} =$ _____ $\dfrac{5}{11} - \dfrac{4}{11} =$ _____

$\dfrac{11}{15} - \dfrac{8}{15} =$ _____ $\dfrac{8}{9} - \dfrac{2}{9} =$ _____ $\dfrac{3}{20} - \dfrac{1}{20} =$ _____ $\dfrac{3}{14} - \dfrac{1}{14} =$ _____

29

$\frac{7}{8} - \frac{3}{8} =$ _____

$\frac{3}{5} - \frac{2}{5} =$ _____

Write each difference in its simplest form.

EXAMPLE:

$\frac{7}{8} - \frac{3}{8} = \frac{7-3}{8} = \frac{4}{8} = \frac{1}{2}$

$\frac{5}{6} - \frac{2}{6} =$ _____

$\frac{4}{6} - \frac{2}{6} =$ _____

$\frac{2}{4} - \frac{1}{4} =$ _____

$\frac{5}{8} - \frac{1}{8} =$ _____

$\frac{5}{6} - \frac{3}{6} =$ _____

$\frac{9}{8} - \frac{4}{8} =$ _____

$\frac{7}{10} - \frac{3}{10} =$ _____

$\frac{7}{16} - \frac{5}{16} =$ _____

$\frac{6}{8} - \frac{5}{8} =$ _____

$\frac{9}{10} - \frac{5}{10} =$ _____

$\frac{3}{5}$ $-\frac{1}{5}$	$\frac{3}{8}$ $-\frac{1}{8}$	$\frac{5}{6}$ $-\frac{1}{6}$	$\frac{7}{10}$ $-\frac{5}{10}$	$\frac{4}{7}$ $-\frac{2}{7}$	$\frac{7}{8}$ $-\frac{2}{8}$
$\frac{3}{4}$ $-\frac{1}{4}$	$\frac{7}{8}$ $-\frac{5}{8}$	$\frac{2}{5}$ $-\frac{1}{5}$	$\frac{7}{8}$ $-\frac{1}{8}$	$\frac{5}{10}$ $-\frac{1}{10}$	$\frac{3}{4}$ $-\frac{2}{4}$
$\frac{7}{10}$ $-\frac{4}{10}$	$\frac{4}{5}$ $-\frac{1}{5}$	$\frac{3}{7}$ $-\frac{2}{7}$	$\frac{5}{10}$ $-\frac{2}{10}$	$\frac{7}{16}$ $-\frac{4}{16}$	$\frac{9}{16}$ $-\frac{5}{16}$
$\frac{5}{8}$ $-\frac{3}{8}$	$\frac{3}{10}$ $-\frac{2}{10}$	$\frac{5}{7}$ $-\frac{4}{7}$	$\frac{4}{6}$ $-\frac{1}{6}$	$\frac{7}{9}$ $-\frac{5}{9}$	$\frac{13}{16}$ $-\frac{5}{16}$

30

$$\frac{7}{8} - \frac{5}{8} = \underline{\qquad}$$

$$\frac{9}{10} - \frac{1}{10} = \underline{\qquad}$$

$$\frac{7}{12} - \frac{1}{12} = \underline{\qquad}$$

$$\frac{8}{9} - \frac{7}{9} = \underline{\qquad}$$

$$\frac{7}{20} - \frac{3}{20} = \underline{\qquad}$$

$$\frac{7}{15} - \frac{2}{15} = \underline{\qquad}$$

e each erence in its plest form.

MPLE:

$$\frac{7}{8}$$

$$\frac{5}{8}$$

$$\frac{2}{8} = \frac{1}{4}$$

$$\frac{9}{25} - \frac{4}{25} = \underline{\qquad}$$

$$\frac{13}{18} - \frac{11}{18} = \underline{\qquad}$$

$$\frac{11}{21} - \frac{2}{21} = \underline{\qquad}$$

$$\frac{5}{14} - \frac{3}{14} = \underline{\qquad}$$

$$\frac{4}{11} - \frac{2}{11} = \underline{\qquad}$$

$$\frac{10}{21} - \frac{4}{21} = \underline{\qquad}$$

$$\frac{7}{24} - \frac{1}{24} = \underline{\qquad}$$

$$\frac{11}{16} - \frac{7}{16} = \underline{\qquad}$$

$$\frac{14}{15} - \frac{2}{15} = \underline{\qquad}$$

$$\frac{12}{25} - \frac{2}{25} = \underline{\qquad}$$

$$\frac{5}{22} - \frac{3}{22} = \underline{\qquad}$$

$$\frac{13}{20} - \frac{7}{20} = \underline{\qquad}$$

$$\frac{11}{12} - \frac{5}{12} = \underline{\qquad}$$

$$\frac{5}{7} - \frac{2}{7} = \underline{\qquad}$$

$$\frac{9}{16} - \frac{3}{16} = \underline{\qquad}$$

$$\frac{17}{18} - \frac{7}{18} = \underline{\qquad}$$

$$\frac{16}{21} - \frac{8}{21} = \underline{\qquad}$$

$$\frac{11}{12} - \frac{1}{12} = \underline{\qquad}$$

$$\frac{9}{14} - \frac{1}{14} = \underline{\qquad}$$

$$\frac{19}{24} - \frac{7}{24} = \underline{\qquad}$$

$$\frac{7}{11} - \frac{2}{11} = \underline{\qquad}$$

$$\frac{9}{22} - \frac{5}{22} = \underline{\qquad}$$

$$\frac{17}{25} - \frac{12}{25} = \underline{\qquad}$$

$$\frac{8}{15} - \frac{7}{15} = \underline{\qquad}$$

$$\frac{15}{16} - \frac{7}{16} = \underline{\qquad}$$

$$\frac{10}{11} - \frac{5}{11} = \underline{\qquad}$$

$$\frac{11}{24} - \frac{5}{24} = \underline{\qquad}$$

$$\frac{11}{13} - \frac{7}{13} = \underline{\qquad}$$

FRACTIONS

elp the Roberts family prepare for Halloween! Add or subtract the fractions. Write your answers in simplest form.

◆　　◆　　◆

WORK SPACE

 Delores offered to decorate 3 cupcakes out of a dozen for Halloween. Anna said she would decorate 5. What part of the dozen did they decorate? _____

 Mr. Roberts put 10 cookies on a cookie sheet to bake. Three-tenths of them burnt and 1/10 of them cracked. What fractional part of the cookies couldn't be used? _____

 Clara handed out candy to 9 trick-or-treaters. She gave candy corn to 3 trick-or-treaters and gave licorice to the rest. To what fractional part of the trick-or-treaters did she give licorice? _____

 Carlos had 10 caramel candies in his bag. He ate 3 of them while trick-or-treating, and 2 of them when he came home. What fractional part of the caramels did Carlos eat? _____

 Mrs. Roberts sent James to buy 8 chocolate bars. On the way home he ate 3 of them and lost 2. What fractional part of the chocolate bars did not make it home with James? _____

 Tom helped Mrs. Roberts hang up 9 decorations. He hung 5 ghosts and 2 pumpkins. What fractional part of the decorations did he hang? _____

 Mr. Roberts took 5 children to buy a pumpkin for Mrs. Roberts. Four of them wanted to buy a big pumpkin, and one of them wanted to buy a small pumpkin. What fractional part of the children wanted to buy a small pumpkin? _____

Continued on page 33.

32

WORK SPACE

Mrs. Roberts needed to make 8 taffy apples for her eight children. She only had enough caramel to coat 5/8 of the apples. What fraction of the apples did she have left to coat? _____

The children ate their Halloween candy. Delores ate 2/5 of hers, and James ate 1/5 of his. How much more did Delores eat than James? _____

After the 8 children ate candy, 3 of them ate so much that they couldn't eat dinner. What fractional part of them ate dinner? _____

The 10 Roberts family members went for a walk after dinner. Three of them counted jack-o-lanterns while the others watched for ghosts. What fractional part of the Roberts family watched for ghosts? _____

Match your answers on pages 32 and 33 to the code. Write the letters in order on the blanks below to solve the riddle.

$$H = \frac{2}{3} \qquad D = \frac{5}{8} \qquad B = \frac{3}{8}$$

$$N = \frac{7}{9} \qquad O = \frac{1}{5} \qquad E = \frac{2}{5}$$

$$Y = \frac{7}{10} \qquad A = \frac{1}{2}$$

Why didn't the skeleton go to the dance?

_ _ _ _ _ _ _ _ _ _ _ _ _ _ !

FRACTIONS

Write each sum in its simplest form.

EXAMPLE:

$5\frac{5}{6}$
$+ 2\frac{5}{6}$

$7\frac{10}{6} = 8\frac{4}{6} = 8\frac{2}{3}$

$5\frac{5}{6}$
$+ 2\frac{5}{6}$

$9\frac{6}{7}$
$+ 4\frac{5}{7}$

$6\frac{5}{8}$
$+ 6\frac{7}{8}$

$5\frac{7}{9}$
$+ 3\frac{5}{9}$

$2\frac{2}{3}$
$+ 3\frac{2}{3}$

$2\frac{3}{4}$
$+ 1\frac{1}{4}$

$5\frac{1}{2}$
$+ 3\frac{1}{2}$

$4\frac{3}{4}$
$+ 2\frac{3}{4}$

$2\frac{3}{5}$
$+ 1\frac{2}{5}$

$3\frac{3}{8}$
$+ 4\frac{1}{8}$

$5\frac{1}{3}$
$+ 1\frac{1}{3}$

$8\frac{1}{5}$
$+ 1\frac{4}{5}$

$1\frac{7}{8}$
$+ 2\frac{3}{8}$

$1\frac{2}{5}$
$+ 2\frac{2}{5}$

$2\frac{4}{7}$
$+ 3\frac{3}{7}$

$4\frac{3}{5}$
$+ 4\frac{4}{5}$

$6\frac{1}{4}$
$+ 1\frac{3}{4}$

$6\frac{1}{8}$
$+ 5\frac{7}{8}$

$4\frac{1}{6}$
$+ 1\frac{5}{6}$

$6\frac{5}{8}$
$+ 4\frac{1}{8}$

$4\frac{1}{3}$
$+ 1\frac{2}{3}$

$8\frac{1}{8}$
$+ 7\frac{5}{8}$

Write each sum in its simplest form.

$$\frac{3}{10} + \frac{1}{10} = \frac{4}{10} = \frac{2}{5}$$

$$4\frac{2}{3}$$
$$+ 1\frac{2}{3}$$
$$\overline{\quad}$$
$$5\frac{4}{3} = 6\frac{1}{3}$$

$\frac{3}{10} + \frac{1}{10} =$ _____ $\frac{3}{12} + \frac{5}{12} =$ _____

$\frac{4}{10} + \frac{3}{10} =$ _____ $\frac{7}{6} + \frac{3}{6} =$ _____

$\frac{1}{3} + \frac{1}{3} =$ _____ $\frac{1}{6} + \frac{1}{6} =$ _____

$\frac{3}{8} + \frac{2}{8} =$ _____ $\frac{1}{5} + \frac{2}{5} =$ _____ $\frac{4}{8} + \frac{5}{8} =$ _____ $\frac{1}{2} + \frac{1}{2} =$ _____

$\frac{1}{4} + \frac{2}{4} =$ _____ $\frac{1}{8} + \frac{4}{8} =$ _____ $\frac{1}{4} + \frac{1}{4} =$ _____ $\frac{2}{4} + \frac{3}{4} =$ _____

$17\frac{2}{6}$ $26\frac{1}{7}$ $19\frac{1}{6}$ $4\frac{2}{3}$ $5\frac{2}{4}$
$+ 9\frac{3}{6}$ $+ 8\frac{2}{7}$ $+ 4\frac{4}{6}$ $+ 1\frac{2}{3}$ $+ 7\frac{3}{4}$

$8\frac{4}{8}$ $9\frac{3}{6}$ $3\frac{1}{3}$ $9\frac{1}{4}$ $4\frac{1}{5}$
$+ 6\frac{5}{8}$ $+ 9\frac{5}{6}$ $+ 5\frac{1}{3}$ $+ 8\frac{1}{4}$ $+ 5\frac{2}{5}$

$7\frac{2}{8}$ $8\frac{1}{2}$ $5\frac{1}{4}$ $4\frac{3}{6}$ $8\frac{5}{8}$
$+ 7\frac{3}{8}$ $+ 7\frac{1}{2}$ $+ 6\frac{3}{4}$ $+ 5\frac{3}{6}$ $+ 8\frac{3}{8}$

35

RIGHT MILLIKEN PUBLISHING CO. MP4074

FRACTIONS AT SCHOOL.

1. On a history test Susan answered $\frac{7}{10}$ of the questions correctly. What fraction of the questions did she miss? _____

2. The fifth grade put on a program. $\frac{2}{9}$ of the students played in the band, and $\frac{6}{9}$ sang in the chorus. What part of the students took part? _____

3. John cut a cake into 12 equal pieces. He ate $\frac{2}{12}$, Kris ate $\frac{2}{12}$, and Tony ate $\frac{2}{12}$. What part of the cake was eaten? _____

4. In Miss Perry's fifth grade class $\frac{4}{7}$ of the students are girls. What part of the students are boys? _____

5. Two-thirds of the students in Central School walk to school. The others ride the bus. What part of the students ride the bus? _____

6. $\frac{5}{8}$ of the students in Dan's class bought lunch on Monday. $\frac{7}{8}$ bought lunch on Tuesday. How many more bought lunch on Tuesday? _____

7. $\frac{3}{5}$ of the cars on the parking lot are four-door cars. The rest are 2-door cars. What part are 2-door cars? _____

8. $\frac{1}{6}$ of the bikes in the bike rack are blue. $\frac{1}{6}$ of the bikes are green. What part are either blue or green? _____

9. $\frac{4}{7}$ of the students chose purple construction paper for their art project. $\frac{1}{7}$ chose brown. What part of the students chose purple or brown? _____

10. In the library $\frac{1}{4}$ of the students chose science fiction stories, $\frac{1}{4}$ chose biographies, and $\frac{1}{4}$ chose adventure stories. The other students did not find a book they liked. What part of the students did not choose a book? _____

◄ **36** ►

LEAST COMMON MULTIPLES.

Write the least common multiple for each pair of numbers.

EXAMPLE: 9 = 3 x 3
12 = 3 x 2 x 2
LCM (9, 12) = 3 x 3 x 2 x 2
LCM (9, 12) = 36

LCM (9,12) =

LCM (6,10) =	LCM (12,16) =	LCM (14,20) =
LCM (8,36) =	LCM (15,27) =	LCM (5,9) =
LCM (11,22) =	LCM (18,24) =	LCM (4,22) =
LCM (15,25) =	LCM (28,35) =	LCM (21,30) =

FRACTIONS

GREATEST COMMON FACTORS.

Write the greatest common factor for each pair of numbers.

EXAMPLE: 18 = 2 x 3 x 3
30 = 2 x 3 x 5
GCF (18, 30) = 2 x 3
GCF (18, 30) = 6

GCF (18,30) =

GCF (15,25) =	GCF (24,32) =	GCF (20,28) =
GCF (27,81) =	GCF (39,52) =	GCF (9,57) =
GCF (12,42) =	GCF (35,49) =	GCF (36,51) =
GCF (32,48) =	GCF (54,66) =	GCF (63,90) =

ADDING FRACTIONS.

Write each sum in its simplest form.

$$+\begin{array}{c} \frac{3}{10} \\ \frac{4}{5} \\ \hline \end{array}$$

$$+\begin{array}{c} \frac{2}{3} \\ \frac{1}{6} \\ \hline \end{array}$$

EXAMPLE:

$$\frac{3}{10} = \frac{3}{10}$$

$$+ \frac{4}{5} = + \frac{8}{10}$$

$$\frac{11}{10} \text{ or } 1\frac{1}{10}$$

$$+\begin{array}{c} \frac{1}{2} \\ \frac{3}{4} \\ \hline \end{array}$$

$$+\begin{array}{c} \frac{7}{8} \\ \frac{1}{4} \\ \hline \end{array}$$

$$+\begin{array}{c} \frac{2}{3} \\ \frac{5}{7} \\ \hline \end{array}$$

$$+\begin{array}{c} \frac{7}{12} \\ \frac{1}{2} \\ \hline \end{array}$$

$$+\begin{array}{c} \frac{1}{6} \\ \frac{4}{9} \\ \hline \end{array}$$

$$+\begin{array}{c} \frac{1}{3} \\ \frac{1}{4} \\ \hline \end{array}$$

$$+\begin{array}{c} \frac{3}{7} \\ \frac{1}{2} \\ \hline \end{array}$$

$$+\begin{array}{c} \frac{3}{5} \\ \frac{1}{3} \\ \hline \end{array}$$

$$+\begin{array}{c} \frac{4}{5} \\ \frac{1}{2} \\ \hline \end{array}$$

$$+\begin{array}{c} \frac{8}{9} \\ \frac{1}{3} \\ \hline \end{array}$$

$$+\begin{array}{c} \frac{3}{8} \\ \frac{1}{3} \\ \hline \end{array}$$

$$+\begin{array}{c} \frac{5}{6} \\ \frac{3}{4} \\ \hline \end{array}$$

$$+\begin{array}{c} \frac{7}{10} \\ \frac{4}{5} \\ \hline \end{array}$$

$$+\begin{array}{c} \frac{1}{2} \\ \frac{7}{9} \\ \hline \end{array}$$

$$+\begin{array}{c} \frac{2}{3} \\ \frac{8}{9} \\ \hline \end{array}$$

$$+\begin{array}{c} \frac{13}{16} \\ \frac{5}{8} \\ \hline \end{array}$$

FRACTIONS

EXAMPLE:

$$\frac{5}{12} = \frac{5}{12}$$
$$+ \frac{5}{6} = + \frac{10}{12}$$
$$\frac{15}{12} = \frac{5}{4} \text{ or } 1\frac{1}{4}$$

Write each sum in its simplest form.

$$\frac{5}{12}$$
$$+ \frac{5}{6}$$
_____

$$\frac{2}{3}$$
$$+ \frac{1}{8}$$
_____

$$\frac{1}{2}$$
$$+ \frac{3}{8}$$
_____

$$\frac{1}{2}$$
$$+ \frac{2}{3}$$
_____

$$\frac{2}{5}$$
$$+ \frac{1}{2}$$
_____

$$\frac{4}{9}$$
$$+ \frac{1}{3}$$
_____

$$\frac{3}{4}$$
$$+ \frac{5}{12}$$
_____

$$\frac{2}{3}$$
$$+ \frac{3}{5}$$
_____

$$\frac{3}{4}$$
$$+ \frac{4}{9}$$
_____

$$\frac{5}{6}$$
$$+ \frac{1}{18}$$
_____

$$\frac{1}{4}$$
$$+ \frac{3}{10}$$
_____

$$\frac{7}{16}$$
$$+ \frac{5}{8}$$
_____

$$\frac{3}{5}$$
$$+ \frac{3}{4}$$
_____

$$\frac{3}{16}$$
$$+ \frac{3}{4}$$
_____

$$\frac{5}{9}$$
$$+ \frac{5}{12}$$
_____

$$\frac{5}{6}$$
$$+ \frac{1}{2}$$
_____

$$\frac{7}{8}$$
$$+ \frac{9}{16}$$
_____

$$\frac{3}{8}$$
$$+ \frac{9}{12}$$
_____

FRACTIONS

ADDING FRACTIONS.

Write each sum in its simplest form.

Row 1:

$$\frac{1}{5}$$
$$\frac{3}{5}$$
$$+ \frac{2}{5}$$

$$\frac{3}{7}$$
$$\frac{4}{7}$$
$$+ \frac{1}{7}$$

$$\frac{5}{12}$$
$$\frac{1}{12}$$
$$+ \frac{7}{12}$$

$$\frac{4}{9}$$
$$\frac{4}{9}$$
$$+ \frac{2}{9}$$

Row 2:

$$\frac{3}{8}$$
$$\frac{5}{8}$$
$$+ \frac{3}{8}$$

$$\frac{3}{4}$$
$$\frac{3}{4}$$
$$+ \frac{1}{4}$$

$$\frac{1}{3}$$
$$\frac{1}{3}$$
$$+ \frac{1}{3}$$

$$\frac{9}{11}$$
$$\frac{10}{11}$$
$$+ \frac{3}{11}$$

$$\frac{1}{6}$$
$$\frac{1}{6}$$
$$+ \frac{1}{6}$$

$$\frac{3}{10}$$
$$\frac{9}{10}$$
$$+ \frac{3}{10}$$

Row 3:

$$\frac{3}{4}$$
$$\frac{1}{8}$$
$$+ \frac{1}{6}$$

$$\frac{4}{9}$$
$$\frac{1}{2}$$
$$+ \frac{2}{3}$$

$$\frac{7}{12}$$
$$\frac{1}{4}$$
$$+ \frac{1}{3}$$

$$\frac{5}{6}$$
$$\frac{4}{9}$$
$$+ \frac{1}{2}$$

$$\frac{4}{5}$$
$$\frac{3}{10}$$
$$+ \frac{5}{6}$$

Row 4:

$$\frac{1}{2}$$
$$\frac{2}{7}$$
$$+ \frac{1}{3}$$

$$\frac{2}{3}$$
$$\frac{3}{5}$$
$$+ \frac{11}{15}$$

$$\frac{7}{20}$$
$$\frac{3}{10}$$
$$+ \frac{1}{4}$$

$$\frac{7}{9}$$
$$\frac{1}{5}$$
$$+ \frac{2}{15}$$

$$\frac{7}{8}$$
$$\frac{3}{4}$$
$$+ \frac{9}{16}$$

FRACTIONS

EXAMPLE:

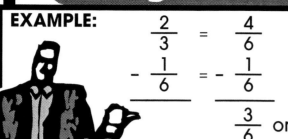

$$\frac{2}{3} = \frac{4}{6}$$
$$-\frac{1}{6} = -\frac{1}{6}$$
$$\frac{3}{6} \text{ or } \frac{1}{2}$$

Write the
difference in its
simplest form.

$$\frac{2}{3}$$
$$-\frac{1}{6}$$

$\frac{4}{5}$ $-\frac{1}{2}$	$\frac{5}{6}$ $-\frac{2}{3}$	$\frac{7}{8}$ $-\frac{3}{4}$
$\frac{5}{7}$ $-\frac{1}{2}$	$\frac{5}{8}$ $-\frac{1}{3}$	$\frac{1}{2}$ $-\frac{2}{5}$
$\frac{2}{3}$ $-\frac{4}{15}$	$\frac{3}{4}$ $-\frac{1}{6}$	$\frac{3}{5}$ $-\frac{3}{8}$
$\frac{7}{8}$ $-\frac{5}{12}$	$\frac{7}{10}$ $-\frac{1}{2}$	$\frac{2}{3}$ $-\frac{1}{4}$
$\frac{2}{3}$ $-\frac{2}{9}$	$\frac{7}{10}$ $-\frac{1}{6}$	$\frac{2}{5}$ $-\frac{1}{4}$

Write the difference in its simplest form.

$$\frac{7}{12}$$
$$-\frac{1}{4}$$
___

$$\frac{1}{2}$$
$$-\frac{1}{3}$$
___

$$\frac{6}{7}$$
$$-\frac{3}{14}$$
___

$$\frac{5}{9}$$
$$-\frac{1}{3}$$
___

EXAMPLE:

$$\frac{7}{12} = \frac{7}{12}$$
$$-\frac{1}{4} = -\frac{3}{12}$$
_____
$$\frac{4}{12} \text{ or } \frac{1}{3}$$

$$\frac{5}{6}$$
$$-\frac{1}{2}$$
___

$$\frac{3}{4}$$
$$-\frac{3}{10}$$
___

$$\frac{8}{15}$$
$$-\frac{2}{5}$$
___

$$\frac{4}{5}$$
$$-\frac{1}{3}$$
___

$$\frac{11}{12}$$
$$-\frac{2}{3}$$
___

$$\frac{7}{9}$$
$$-\frac{1}{2}$$
___

$$\frac{1}{2}$$
$$-\frac{3}{10}$$
___

$$\frac{2}{3}$$
$$-\frac{2}{7}$$
___

$$\frac{3}{4}$$
$$-\frac{2}{3}$$
___

$$\frac{5}{12}$$
$$-\frac{1}{6}$$
___

$$\frac{11}{14}$$
$$-\frac{2}{7}$$
___

$$\frac{7}{12}$$
$$-\frac{1}{3}$$
___

FRACTIONS

FINDING THE SIMPLEST FORM.

Write each sum in its simplest form.

$\dfrac{2}{6}$ $+ \dfrac{3}{10}$	$\dfrac{3}{8}$ $+ \dfrac{1}{2}$	$\dfrac{3}{10}$ $+ \dfrac{1}{5}$	$\dfrac{2}{4}$ $+ \dfrac{3}{5}$	$\dfrac{2}{3}$ $+ \dfrac{4}{8}$
$\dfrac{7}{9}$ $+ \dfrac{2}{18}$	$\dfrac{4}{6}$ $+ \dfrac{3}{4}$	$\dfrac{2}{5}$ $+ \dfrac{3}{6}$	$\dfrac{4}{8}$ $+ \dfrac{1}{2}$	$\dfrac{4}{16}$ $+ \dfrac{2}{8}$

Write each difference in its simplest form.

$\dfrac{3}{5}$ $- \dfrac{1}{2}$	$\dfrac{6}{8}$ $- \dfrac{1}{2}$	$\dfrac{6}{10}$ $- \dfrac{1}{5}$	$\dfrac{2}{3}$ $- \dfrac{1}{2}$	$\dfrac{3}{8}$ $- \dfrac{1}{10}$
$\dfrac{5}{6}$ $- \dfrac{2}{3}$	$\dfrac{5}{12}$ $- \dfrac{2}{8}$	$\dfrac{4}{5}$ $- \dfrac{2}{4}$	$\dfrac{2}{3}$ $- \dfrac{1}{8}$	$\dfrac{4}{8}$ $- \dfrac{1}{6}$
$\dfrac{6}{7}$ $- \dfrac{2}{14}$	$\dfrac{3}{9}$ $- \dfrac{2}{18}$	$\dfrac{3}{4}$ $- \dfrac{1}{5}$	$\dfrac{7}{10}$ $- \dfrac{1}{2}$	$\dfrac{2}{3}$ $- \dfrac{1}{4}$

44